D1134721

Environmental Health and Housing

With significant amounts of former council run housing returning to the private sector, and policy continuing to favour this sector in housing provision, the need for expertise in private sector environmental health enforcement is greater than ever.

Environmental Health and Housing will provide both students and professionals with comprehensive coverage of issues related to both social and private housing. The book includes basic technical information for completing house surveys, detailed yet clear backgrounds to and explanations of applying relevant legislation, and discussion of current policy and strategy. All this is backed up with case studies and examples of how theory and law are put into practice in real situations.

Unique in its coverage, clearly illustrated and covering such diverse topics as housing defects, caravan sites, asylum seekers and social exclusion, *Environmental Health and Housing* is an essential purchase for all students and professionals in the housing sector.

Jill Stewart, Senior Lecturer in Environmental Health and Housing, University of Greenwich

Clay's Library of Health and the Environment

Environmental Health and Housing

Clay's Library of Health and the Environment: Volume 1

Jill Stewart

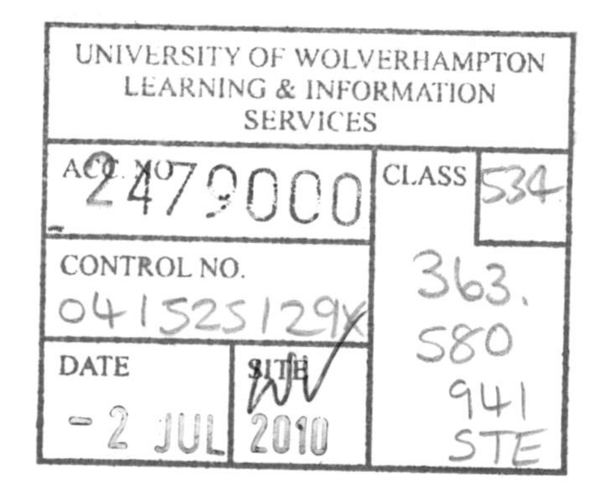

Taylor & Francis
Taylor & Francis Group
LONDON AND NEW YORK

First published 2001 by Taylor & Francis
2 Park Square, Milton Park, Abingdon, Oxon, OX14 4RN

Simultaneously published in the USA and Canada
by Taylor & Francis
270 Madison Ave, New York NY 10016

Taylor & Francis is an imprint of the Taylor & Francis Group

Transferred to Digital Printing 2008

Typeset in Palatino by Wearset, Boldon, Tyne and Wear

British Library Cataloguing in Publication Data
A catalogue record for this book is available from the British Library

Library of Congress Cataloging in Publication Data

Stewart, Jill, 1967–
Environmental health and housing/Jill Stewart.
p. cm – (Clay's library of health and the environment; v. 1)
Includes bibliographical references and index.
ISBN 0-415-25129-X (alk. paper) –
ISBN 0-415-25130-3 (pbk. : alk.paper)
1. Housing – Environmental aspects – Great Britain. 2. Public housing – Environmental aspects – Great Britain. 3. Housing and health – Great Britain. 4. Environmental health – Great Britain. I. Title. II. Series.

HD7333.A3 S74 2001
363.5'0941 – dc21 00-067084

Publisher's Note
The publisher has gone to great lengths to ensure the quality of this reprint but points out that some imperfections in the original may be apparent.

This book is dedicated to
Albert James Carey
and
Evelyn Gladys McBain
whose lives and families taught me the value of decent, secure and affordable housing.

Contents

Preface

The private housing sector comprises the majority tenure in the UK, and to a large extent it regulates itself because owners are willing and able to invest in what amounts to a substantial capital asset. However, the nation's housing stock is ageing and requires constant maintenance to keep it in a reasonable state of repair. Many house owners are unable, or sometimes unwilling, to maintain their accommodation in a reasonable state of repair. There are also those who may require assistance to make their homes suitable for their specific needs, including the growing elderly population and those with disabilities. Many lower-income groups have little active choice in the housing market, and, unable to access social housing (as defined in Chapter 1), occupy accommodation at the bottom end of the private rented sector where conditions can be extremely poor.

While there is a plethora of information concerning social housing, there is less written about private sector housing, and what is written tends to be disparate and some housing law and practice texts require updating. The sector is important in the housing cycle – many rent before buying, some are never in a position to buy, others do not wish to, some move from private renting into the social sector, and so on. There is also little written about those who live in structures other than traditional housing, including mobile homes, caravans, houseboats and self-built structures where security of tenure is closely tied to land rights. The sector itself is complex, with a complex structure of associated housing and housing-related law and practice. There is no one book dedicated to drawing the complexities of the private housing sector together and consolidating the range of knowledge required by those with an interest in the sector.

This book seeks to fill that gap. It provides, in one volume, the range of housing knowledge needed to understand issues in the sector, its threats and opportunities. The book provides a technical overview of housing, from an introduction to identifying housing defects and terminology to drafting specifications of work, and it provides an understanding of tenancy legislation and basic housing finance. It provides a background to the history of private sector housing renewal, housing and health, inequality in housing, domestic energy efficiency and other related issues. It consolidates current and proposed legislation to address poor and unsuitable living conditions, before considering how this can be incorporated into local housing strategies to make best use of available resources.

The book seeks to provide a complete 'why' and 'how to' in dealing with conditions in the private housing sector. By its nature, this involves not only just consideration of housing and living conditions, but also gives an insight into residents, why they live there and conditions of tenancy, which are all relevant in tackling poor housing and living conditions.

The book is designed to appeal to a wide housing audience, including environmental health staff, and housing officers who increasingly have a role to play in the private housing sector, particularly private rented housing. It seeks to provide guidance on regulating standards and identifying increased housing potential in the private sector, whether formally or informally, and to provide further information for those who simply wish to find out a bit more. It can be used as an initial training manual or as a source of updating information in key areas, and is divided into chapters that stand-alone and are easy to dip in to. It seeks to provide an 'at-a-glance guide' and reference manual for those new to private sector housing law, an *aide-mémoire* to those already involved and an easy-to-read reference book to those active in associated issues in the private housing sector. It includes many photographs of common conditions found in the sector and typical case studies that housing staff in local authorities regularly face, illustrating how housing theory relates to housing practice.

The format of this book is as follows, with a contents outline at the start of each chapter. Chapter 1 introduces the book. Chapter 2 provides a background to legal conditions. Chapter 3 introduces essential basis knowledge required as a background to deal with poor housing conditions. Chapter 4 looks specifically at the environmental health housing law and its application. Chapter 5 considers issues involved in, and the development and implementation of, housing strategies. Chapter 6 offers some conclusions.

Housing is a rapidly changing policy area, and the time this book comes to print there is likely to be a new Housing Act. However, it is written with this in mind and has highlighted the current issues under consideration in the Green Paper, *Quality and Choice: A Decent Home for All*, published in April 2000.

Jill Stewart

Acknowledgements

First and foremost, I am grateful to W. H. Bassett (Editor of *Clay's Library of Health and the Environment*) and Michael Doggwiler (Senior Editor, Spon Press – an imprint of the Taylor and Francis Group) for their continued and enthusiastic advice, ideas, guidance and support throughout the compilation of this book.

I am indebted to Three Rivers District Council for permission to use the photographs and paperwork such as survey sheets, risk inspection sheets, etc.; to the Chartered Institute of Environmental Health for permission to use the picture of Edwin Chadwick in Section 2.1; to Phil Parnham of Sheffield Hallam University, the artist of the useful 'TRIS' diagrams in Sections 3.1 and 3.3. Other acknowledgements are cited where relevant in the body of the text.

Some of the text relating to health and inequality in housing (Chapter 2) first appeared in *Journal of the Royal Society of Health*, 119(4), and is cited in the references and further reading sections as appropriate.

I would also like to thank my colleagues at the School of Land and Construction Management, University of Greenwich, in particular Paul Balchin and Maureen Rhoden for their advice and help, but also Stuart Allen, Fiona Bushell and Veronica Habgood for their continued support. Thanks are also due to my past and present colleagues in environmental health and housing across many statutory and non-statutory organisations from whom I have learned how best to understand and apply housing law and practice over more than a decade.

Finally, but by no means least, I also thank my mother, Janet, for proof-reading the final text, and my husband, James, who supported me through it.

Abbreviations

AFD Automatic Fire Detection
AMA Association of Metropolitan Authorities
ACG Annual Capital Guideline
BABIE Bed and Breakfast Information Exchange
BCA Basic Credit Approval
BRE Building Research Establishment
BSC Boat Safety Certificate
CBR Campaign for Bedsit Rights
CIEH Chartered Institute of Environmental Health
CIH Chartered Institute of Housing
CPG Common Parts Grant
CRI Capital Receipts Initiative
DAN Deferred Action Notice
DETR Department of the Environment, Transport and the Regions
DFG Disabled Facilities Grant
DoE Department of the Environment
DoH Department of Health
DSS Department of Social Security
ECA Energy Conservation Act
EHA Empty Homes Agency
EHCS English House Condition Survey
EHO Environmental Health Officer
EU European Union
FFE Fire Fighting Equipment
FMO Flat in Multiple Occupation
GIA General Improvement Area
GRS Group Repair Scheme
HAA Housing Action Area
HECA Home Energy Conservation Act
HEES Home Energy Efficiency Scheme
HIP Housing Investment Programme
HMO House in Multiple Occupation

HRA	Home Repairs Assistance
HSE	Health and Safety Executive
JRF	Joseph Rowntree Foundation
LGA	Local Government Association
LGMB	Local Government Management Board
LOTS	Living over the shops
LRC	London Research Centre
MSCA	Most Satisfactory Course of Action
NASS	National Asylum Support Service
NEA	National Energy Agency
NFHA	National Federation of Housing Associations
NPV	Net Present Value
NRA	Neighbourhood Renewal Assessment
PHA	Public Health Association (USA)
PRS	Private Rented Sector
PSRSG	Private Sector Renewal Support Grant
RTIA	Rate Taken into Account
SAP	Standard Assessment Procedure
SCA	Supplementary Credit Approval
SCG	Specified Capital Grant
SRB	Single Regeneration Budget
TRIS	Tenants Resource and Information Service
WHO	World Health Organisation

Chapter 1

Introduction

Housing is about people's lives. People's homes and their local communities can influence health and well-being, community and family ties, and access to wider society. It then follows that decent housing has a positive effect on a person's lifestyle and their life chances. Those living in decent housing are more likely to maintain good health, access to education, health, community facilities and employment opportunities. Conversely, those who live in poor housing, particularly within deprived communities, are less likely to achieve such benefits. Poor housing can cause or aggravate existing physical or mental ill-health, or existing housing may not be suitable to meet someone's changing needs for a variety of reasons, such as age or disability. Decent and appropriate housing supply, with appropriate community services and support, is the cornerstone to a socially inclusive society.

Much is written about social housing, its purpose, function, management and so on. There is less written about the private sector, despite it comprising the vast majority of housing stock, including many lower income households traditionally housed in the social sector. Radical changes in recent years have seen a major ideological shift from local authorities as housing providers to housing enablers, and part of this shift has involved looking at the private sector in a new light. It offers opportunities as well as threats.

This book, therefore, seeks to fill a gap in literature on the private housing sector. For the purposes of this book, the private housing sector is defined to include owner-occupation, leaseholders and private sector tenants where the landlord's purpose for letting is normally as a business, profit-making enterprise. It also includes residences that do not fall into the social housing category, such as travellers' accommodation, self-build structures, caravans and mobile homes, as well as house-boats, where accommodation is very much tied up with land ownership rights. The social housing sector is here seen to include local authority, housing association and not-for-profit housing supply organisations, which have access to alternative funding arrangements. This immediately makes dealing with the private sector fundamentally different, because the sector has different objectives, ownership features and other interests. This necessitates a sound understanding of relevant housing law and practice.

Many organisations have a role to play in housing standards and management. Statutory organisations include the Department of the Environment, Transport and the Regions (DETR) and local authorities, and recent years have seen a growth in alternative housing suppliers, such as housing associations, which are increasingly contracting to local authorities to provide social housing. The Chartered Institute of Housing (CIH) and the Chartered Institute of Environmental Health (CIEH) are the national independent professional organisations concerned with promoting housing standards across all tenures, but the CIH is more commonly associated with social housing, and the CIEH with private sector housing. Interest in housing also comes from other sectors, in particular from nationally recognised pressure groups, including Shelter and the Campaign for Bedsit Rights (CBR), that provide an important role in promoting decent housing conditions as well as a wealth of valuable information. A summary of key housing organisations is included in Table 1.1.

Private sector housing: the local authority role

The fundamental purpose of local authority environmental health activity is to break the link between poor housing and ill-health, through a mixture of enforcement and enabling provisions. The objectives include repair and/or improvement of individual houses, rehabilitation on an area basis, a wider strategic role in anti-poverty, social exclusion and political initiatives as well as the increasing use of private sector as 'alternative' to social housing.

Local authority housing functions are vested in several pieces of legislation, which contain both statutory duties and discretionary powers. A summary of housing and housing-related legislation delivered by local authorities is given in Table 1.2. Statutory obligations are defined under legislation as 'duties' and must be delivered. An example is taking action to deal with a statutorily unfit house under the Housing Act 1985 (as amended). Where legislation states that a local authority 'may' take action, this action is discretionary. This means that local authorities are given the power, but not the duty, to act under legislation. Each local authority must determine policy, by formal committee decision, to deal with issues such as discretionary renovation grants under the Housing Grants, Construction and Regeneration Act 1996.

There is an increasing variety of personnel involved in addressing conditions in the private sector. Traditionally, environmental health officers (EHOs) were the key professionals in local government with legal powers to deal with private sector housing conditions. Policy changes favouring the private sector mean that more people are now involved in identifying private sector properties and making them available for rent, such as housing and homeless officers who play a more enabling role. Such officers now frequently work closely together in an attempt to secure an increasing supply of housing through the private rented sector using a range of new initiatives. While traditional housing enforcement activities as still very prevalent, and fundamental, a new role is emerging in the way this sector's use is being encouraged to meet housing need.

What has also changed is the way in which local authorities use their enforcement powers. There is now much more emphasis on client involvement in their own housing, and having their own choices, where this is possible, in terms of housing conditions. This change has been part of a wider change in issues such as customer service and quality initiatives, although, of course, it can be more difficult in private sector housing, particularly when dealing with landlords. What is important is that those vested with an enforcement or advisory interest in private sector housing use the legislation and resource opportunities available to them sensitively to achieve a viable outcome in any given situation.

Despite the move to the private rented housing and the recognition of part of this sector's poor conditions, there has been reduced capital expenditure in this sector. The 1996 English House Condition Survey illustrated that poor housing is frequently associated with low income (DETR 1998). The private rented sector proportionally comprises the poorest sector of housing, particularly in houses in multiple occupation, yet government ideology during the 1980s and 1990s sought to increase this sector's supply. The Audit Commission Report on Healthy Housing (Audit Commission 1991) noted that despite considerable staff resources, only a relatively small proportion of private sector properties in poor condition had been subject to environmental health enforcement powers, despite being frequently occupied by the most disadvantaged members of society.

Perhaps as a response to this, but also seen across other public sector services, there has been increased central guidance on housing action at local level. This includes the key document, Department of the Environment *Circular 17/96, Private Sector Renewal: A Strategic Approach* (DoE 1996). Such guidance encourages the use of private sector stock in innovative and proactive ways, such as making greater use of vacant properties and encouraging domestic energy efficiency. The government is currently seeking consultation on its Green Paper, *Quality and Choice: A Decent Home for All*, published in April 2000, to bring forward new housing initiatives, with a renewed emphasis on developing and delivering local strategies that meet the local need (DETR 2000). The government's current housing objectives are set out in Table 1.3.

Table 1.1 Housing organisation contact details

Organisation and address	*Function of organisation*	*Contact details*	*Regular publications*
Building Research Establishment, Garston, Watford WD2 7JR	UK's leading centre for construction and fire expertise and consultancy. BRECSU researches and publishes independent advice on energy efficiency in buildings on behalf of the government	http://www.bre.co.uk	*Constructing the Future* (quarterly) – reviews current BRE research
Campaign for Bedsit Rights, 88 Old Street, London EC1V 9HU	Specialist unit within Shelter to carry out campaigning, policy and information initiatives on issues affecting the private rented sector; aims to improve the rights and living conditions for tenants in HMO-type accommodation	http://www.shelter.org.uk	Various leaflets, research, good practice guidance and bulletins
Centre for Housing Policy, York	Established in 1990 as a centre of excellence in policy relevant to housing research	http://www.york.ac.uk/inst/chp	Summaries of past and on-going housing research
Chartered Institute of Environmental Health, Chadwick Court, 15 Hatfields, London SE1 8DJ	Independent professional body representing the interests of environmental health professionals through promotion of professional, educational and membership services	http://www.cieh.org.uk	*Environmental Health Journal* (monthly), *Environmental Health News* (weekly), *Health and Housing Insight*
Chartered Institute of Housing, Octavia House, Westwood House, Coventry CV4 8JP	Independent professional body to promote and encourage the provision and management of affordable housing for all	http://www.cih.org	*Housing* (monthly), *Inside Housing* (weekly), http://www.atlas.co.uk/inside/

Department of the Environment, Transport and the Regions, Eland House, Bressenden Place, London SW1E 5DU	Government department responsible for housing policy. Publishes free summaries of completed DETR housing research projects and information leaflets on a range of housing topics	Home page: http://www.detr.gov.uk Housing page: http://www.housing.detr.gov.uk	*Housing Signpost – A Guide to Research and Statistics* DETR information leaflets; DETR Free Literature, PO Box 236, Wetherby, West Yorkshire LS23 7NB
Empty Homes Agency, 195–197 Victoria Street, London SW1E 5NE	Established in1992 to highlight the waste of empty English homes and to devise solutions and good practice to tackle empty properties	http://www.emptyhomes.com eha@globalnet.co.uk	
Joseph Rowntree Foundation, The Homestead, 40 Water End, York YO30 6WP	Major source of housing and social policy research	http://www.jrf.org.uk	*Findings*
Local Government Association	Established in 1997 to represent local government and to promote democratic local communities	http://www.lga.gov.uk	
London Research Centre	Provider of research and information on urban affairs – recently absorbed into the new Greater London Authority	http://www.london-research.gov.uk	*Research News*, *Housing and Social Research*
National Energy Action Charity, National Office, St Andrew's House, 90–92 Pilgrim Street, Newcastle-upon-Tyne NE1 6SG	Established in 1981 to campaign for warm homes and an end to UK fuel poverty – has assisted and advised many low-income households	http://www.nea.org.uk info@nea.org.uk	
National Housing Federation	National body representing the independent social housing sector	http://www.housing.org.uk	*Housing Today* http://www.housing.today.org.uk
Shelter, 88 Old Street, London EC1V 9HU	National organisation to improve the lives of the homeless and badly housed people	http://www.shelter.org.uk info@shelter.org.uk	*Roof*

Table 1.2 Summary of key current legislation concerning living conditions

Act	*Features*
Public Health Act 1936	Provisions for dealing with insanitary housing and some miscellaneous housing provisions
Building Act 1984	Contains accelerated procedures for dealing with statutory nuisance; provides remedy for dilapidated dwellings
Caravan Sites and Control of Development Act 1960	Governs caravan site licencing conditions and other controls
Housing Act 1985 (as amended)	Consolidating Act containing duty to consider annually local housing conditions; provides the statutory standards of fitness and overcrowding as well as the discretionary powers for disrepair and controls for HMOs
Environmental Protection Act 1990	Contains statutory nuisance provisions to deal with 'premises'
Criminal Justice and Public Order Act 1994	Removed the local authority duty to provide gypsy sites and strengthened powers to deal with unauthorised encampments
Home Energy Conservation Act 1995 and Energy Conservation Act 1996	New local authority duty to reduce domestic energy emissions to cut greenhouse gases
Housing Act 1996	Provides powers in respect of HMOs, in particular for registration schemes and an anticipated HMO Code of Practice
Housing Grants, Construction and Regeneration Act 1996	Provides for discretionary grant assistance for renovation works and mandatory grant assistance for disabled facilities
Asylum and Immigration Act 1999	Removed asylum seekers from mainstream welfare support and replaced this with a separate system of cashless support and new provision for housing
Current Green Paper *Quality and Choice: A Decent Home for All* (2000)	The government is currently considering, and inviting comment, on the following areas: • Developing a housing strategy for the twenty-first century. • Challenges faced. • Making it work locally. • Encouraging sustainable home ownership. • Promoting a healthy private rented sector. • Reforming social housing for the twenty-first century. • Raising the quality of social housing. • Providing new affordable housing. • Choice in social housing. • Moving to a fairer system of affordable rents. • Improving housing benefit. • Tackling other forms of housing-related social exclusion.

Source: DETR (2000)

Table 1.3 Government's key principles for housing policy

- Offering opportunity, choice and a stake in their home regardless of tenure
- Ensuring adequate housing supply to meet need
- Giving individuals responsibility to provide their own housing where possible, and help to those who cannot
- Improving the quality and design of new housing stock and environments, moving toward an urban renaissance and protecting the countryside
- Delivering modern, efficient, secure customer-focused services and empowering individuals to influence them
- Reducing barriers to work, especially in respect of benefit and rent policy
- Supporting vulnerable people and tackling forms of social exclusion including poor housing, homelessness, poverty, crime and poor health
- Promoting sustainable development to support thriving, balanced communities and quality of life in urban and rural areas.

Source: based on DETR (2000)

Much of the Green Paper focuses on the social sector and as such is outside the scope of this book. In terms of the private housing sector, the key issues under consideration include:

- A continued shift toward sustainable owner-occupation encouraged through fiscal incentives.
- Increased flexibility in the renovation grants system, particularly to provide more effective help to low-income owners of poor-quality housing and consideration of whether to retain the existing grant scheme or to introduce a general power providing local discretion to determine grant eligibility, conditions and amount.
- Emphasis on tackling concentrations of poor housing through area-based renewal schemes.
- Exploration of alternative sources of funding housing renewal, including greater use of equity and a move toward loans and maintenance services and new ways of levering in private sector finance.
- Exploration of increased flexibility on area renewal conditions and criteria.
- Promoting the role of the private rented sector through encouraging the better landlords, while tackling the worst, e.g. through voluntary accreditation schemes, discretionary licencing powers and possible conditions for housing benefit payments, with closer regulation by the Rent Service and a new role for registered social landlords to manage private sector properties.
- Proposals to introduce a new housing health and safety rating system (this is examined in more detail in Section 4.2).
- Introduction of a compulsory HMO licencing system to modernise and rationalise current controls, it being an offence to operate an

HMO without a licence or being in breach of licencing conditions (this is examined in more detail in Section 4.8).

- Proposals to provide for discretionary licencing of properties and landlords in problematic properties or neighbourhoods, now predominated by the bottom end of the private rented sector and often associated with criminal activity.
- A general proposal to improve the housing benefit system to reduce bureaucracy, complexity and fraud, to increase the incentive and ability to work, and to explore alternatives to complement housing policy, such as single room rent.

There are, of course, further consultative and administrative stages until the current Green Paper becomes legislation, and at the time of writing it is too early to comment on in detail – further information is included as necessary in the body of the text.

Chapter 2

Background to legal conditions

This chapter provides a background to legal conditions. It traces the history of private sector housing renewal through to current concerns with sustainability; discusses issues in housing and health; explores the role of the private rented sector; investigates inequality in housing and homelessness; illustrates the debate surrounding housing asylum seekers; and overviews current issues in fuel poverty and energy efficiency.

The chapter is presented as follows:

2.1 History of private sector housing renewal
2.2 Housing and health
2.3 Role of the private rented sector
2.4 Inequality and poor housing conditions
2.5 Homelessness
2.6 Asylum seekers and housing
2.7 Fuel poverty, energy efficiency and conservation

2.1 History of private sector housing renewal

Outline

A background knowledge in the development of private sector housing renewal policies is important in informing an understanding of current housing conditions, law and practice. It is also useful to overview which policies were successful and why, and, conversely, which policies were not so successful, and why. A comprehensive, honest and sensitive overview is useful in developing sustainable new policies. This section, which can only offer a summary of key features and dates, traces the history of private sector housing renewal to its roots in the Victorian public health movement, changes in policy since then, to where it is at now. It seeks to show how much housing policy relates to wider social change and political ideology, and briefly considers the development of social housing, and is summarised in Table 2.1.

Table 2.1 History of housing law

Key dates and features	*Committees and legislation*	*Comments*
1840s – Industrial Revolution and urbanisation, aggravating poor housing. *Laissez-faire* ideology. Birth of public health and philanthropic activity in housing.	Public Health Act 1848 Common Lodging Houses Act 1851 Labouring Classes Lodging Houses Act 1851 (Lord Shaftesbury) Torrens Acts 1868 and 1879	Introduction of legislation empowering local authorities to deal with poor housing conditions
Late 1800s	Artisans and Labourers Dwellings Act 1868 Artisans and Labourers Dwellings Improvement Act 1875 (Cross Acts)	Provided for local authority intervention in housing conditions and allowed for rates for local authority building
	Public Health Act 1875	Local authority powers to make building by-laws
	Royal Commission on Housing of the Working Classes Housing of the Working Classes Act 1890 Town and Country Planning Act 1909	Provided a formal structure to *ad hoc* housing schemes and encouraged further schemes
1914–18 – Predominantly privately rented housing, low incomes, rent strikes, massive social change	Tudor Walters Committee Rent and Mortgage Restriction Act 1915 Housing (Additional Powers) Act 1919 The Housing and Town Planning Act 1919 (Addison Act) Housing Act 1923 (Chamberlain Act)	Rent controls introduced 'Homes Fit for Heroes' to standards determined by Tudor Walters Committee with a subsidy for new houses, extended in 1923 with Trade Union support. New duty to survey district's housing needs
1920s–1930s – Growth of Building Societies, development of suburbia, increasing owner occupation and decline of private rented sector, self-help schemes, etc., alongside growing depression and poverty; national debt, and socio-political unrest, worsening of housing conditions e.g. overcrowding and lack of renewal. Health Minister Sir Hilton Young	Housing (Financial Provisions) Act 1924 (Wheatley Act)	Long-term house building programme
	Housing Act 1930 (Greenwood Act)	Power to grant rent rebate
	Housing Act 1935	Definition of overcrowding

Table 2.1 continued

Key dates and features	*Committees and legislation*	*Comments*
promised £95 million on slum clearance and rehousing in 1933, combining employment with social provision		
1939–45 – 3.5 million homes damaged in air-raids, 173,500 slum houses from the 1930s' programmes still lived in; use of unfit housing to meet need; poor housing still common	1944 – Dudley Committee	Recommended standards for local authority housing
Post-1945 – Introduction of the Welfare State. Development of New Towns, but not until 1955 that housing renewal was in full swing	New Towns Act 1946 Town and Country Planning Act 1947 Housing Act 1949	New Town Development Corporations Development rights consolidated Grants for standard amenities introduced
1950s–1960s – Continued decline in private rented sector; increase in owner occupation. Welfare State remained a priority with full employment. Growth of pressure groups together with Rackman legacy and *Cathy Come Home* had a huge impact on social attitudes on housing and homelessness. Rent officers established and birth of Housing Corporation. Tale ends of mass clearance and area municipalisation	Housing Act 1952 Housing Repairs and Rents Act 1954 Housing Subsidies Act 1956 Rent Act 1957 Housing Act 1957 Housing Act 1961 1961 – Parker Morris Committee Housing Act 1964 Protection from Eviction Act 1964 1965 – Milner Holland Report (Housing in Greater London) Rent Act 1965 Housing Subsidies Act 1967 Housing Act 1969	Housing subsidy increased Slum clearance High-rise building encouraged Fitness standard consolidated Controls for HMOs introduced. Subsidy costs based on Parker Morris standards New powers for repair, GIAs introduced
1970s – Social unrest; influence of Europe on domestic policy. Ombudsman service introduced in 1974. Recognition of early failure of some municipal estates	Housing Act 1974 Local Government Act 1974 Housing (Homeless Persons) Act 1977	HAAs introduced. More interventionist local authority role through grants and CPO Right to permanent housing for the priority homeless
Recent – recognition of social exclusion, emphasis	Housing Act 1980	Right to Buy, tenants charter, amendments to finance

Table 2.1 continued

Key dates and features	*Committees and legislation*	*Comments*
on partnerships and commissioning, sustainability. Increased emphasis on local housing strategy. Rise of the 'New Left'	Housing Act 1988	Consolidating Act.
	Housing Act 1988	Introduction of deregulated tenancies (Assured), Housing Action Trusts, Tenants Choice, mixed funding
	Local Government and Housing Act 1989	Mandatory means tested grants linked to fitness standard; ring-fenced housing revenue account
	Leasehold Reform, Housing and Urban Development Act 1993	CCT in local authority housing
	Housing Grants, Construction and Regeneration Act 1996	All grants discretionary, except DFGs
	Housing Act 1996	Changes to homeless, enforcement processes etc.

Public health pioneers

The Industrial Revolution had seen thousands of people move away from rural life and into the emerging polluted and overcrowded cities to find work. The massive influx from country to town put huge pressure on *urban* areas. Edwin Chadwick, the father of the environmental health profession in the Victorian era, first made the link between health and housing (Figure 2.1). In 1842 he presented a report on the Sanitary Conditions of the Labouring Population of Great Britain to Parliament (cited in Brown and Savage 1998, Hibbert 1988). It was the first time anyone had linked living conditions to health and revealed a desperate picture of overcrowded, damp, unventilated houses lacking adequate drainage and proper water supplies. Chadwick believed that such conditions caused and aggravated ill health among the working classes, resulting in an average life expectancy of an urban working class person of 12–15 years. Engels' 1844 Condition of the Working Classes (cited in Briggs 1987, Hibbert 1988) painted a similar picture.

Pioneered by Chadwick, the Public Health Act 1848 had attracted much criticism, but by now there was no turning back and the General Board of Health was born. This early legislation focused around public health rather than housing conditions, mainly concentrating on issues such as drainage and water supply, possibly because recent cholera outbreaks had affected all social classes, not just the poor. The Victorian attitude was very much *laissez-faire*; the poor were seen to be responsible for

Figure 2.1 Edwin Chadwick in 1848

their own lot, and it was up to them to take steps to counteract their housing conditions, where overcrowding, disrepair, lack of facilities and so on were rife. Providing the poor with decent housing was seen as nothing short of revolutionary, and certainly not a government function.

Chadwick was not the only one changing attitudes in the Victorian era. Octavia Hill, pioneering in her time for being both a philanthropist and a woman set on social change, established the first social housing at Paradise Place, Marylebone in London. Other wealthy families committed to social change also became involved. Joseph Rowntree (1805–59), for example, of a Quaker family, believed that social change was possible and he began to challenge the Victorian notion that poverty was a natural, and therefore acceptable, state of affairs – the Joseph Rowntree Foundation is still a key institute for housing and social research. Charles Booth's Life and Labour of the People of London from 1899 to 1903 developed a new attitude toward poverty, which was increasingly seen as a social phenomenon rather than as a fact of life (cited in Briggs 1987, Hibbert 1988).

Many low-income households had to share accommodation, so common lodging houses became popular. Lord Shaftesbury (previously known as Lord Ashley) pioneered two Acts to tackle such poor conditions. The Common Lodging Houses Act 1851 introduced some controls to such accommodation and the Labouring Classes Lodging Houses Act 1851 enabled local authorities to create lodging houses to reduce homelessness, although these Acts had little impact on housing conditions due to a lack of associated resources.

However, the continued pressure for legal improvements had gained momentum. The Torrens Acts 1868 and 1879 enabled local authorities to deal with individual insanitary houses, although not areas of bad housing. The first legal powers for area action were introduced in private legislation in Manchester in 1867. With little security of tenure, tenants were readily evicted when an owner wished to sell the property for commercial interests. This changed with the Cross Acts (Artisans and Labourers Dwellings Improvement Acts) 1875 and 1879, which allowed local authorities to intervene in unfit housing, to clear and redevelop land for the purpose of improvement, through designated schemes, for the working classes, with provision for Compulsory Purchase. The simultaneous Public Health Act 1875 enabled proactive local authorities to adopt by-laws to control building standards.

One problem for local authorities with these new Acts was that of cost. Local authorities were reluctant to spend funds on tackling working-class housing conditions but public pressure continued, leading to the Artisans Dwelling Act 1882, which provided closely controlled financial assistance from central government. Progress remained slow and a Royal Commission was established in 1884 to consider housing for the working classes. It was made up of respected public figures such as Lord Shaftesbury and Chadwick, who was by then President for the Association of Sanitary Inspectors. The Commission's work led to the Housing of the Working Classes Act 1885, which required local authorities to achieve proper sanitary conditions of housing in their areas through an implied condition that houses would be fit at the commencement of the holding, provided for by-laws to deal with houses let as lodgings and required the supervision of tents and vans used for dwellings (Foskett 1999). Simultaneously, philanthropic charitable trusts were providing some working-class housing and were developing new forms of housing management. Hill, in particular, pioneered early tenant involvement in housing management, largely through attempting to encourage certain behaviour .

The Housing of the Working Classes Act 1890 drew together the miscellaneous *ad hoc* housing schemes in an attempt to consolidate best policy and to encourage housing where none currently existed. The Act provided for dealing with unhealthy areas and improvement schemes, unfit dwelling houses and powers to provide lodging houses, which provided the administrative framework for later housing measures. However, it came with no funding and local authorities were expected to raise the money themselves. The Housing and Town Planning Act 1909 enabled local authorities powers to control development and it introduced some controls on housing development and design, such as prohibiting back-to-back houses, and it recognised the role of building societies in housing. As a result, prior to the First World War, less than 1 per cent of housing stock had been provided by municipal and philanthropic activity.

Regional differences remained in both legislation and policy. Generally, large northern industrial towns were developing long-term plans for housing redevelopment. By 1894, the London Building Act had legislated for the first building code and building standards requiring minimum standards of sanitation and layout. The idea of the garden city and the new town was born at the turn of the century, together with the notion of fitness for human habitation, remaining a precedent for local authorities planning and control powers in respect of housing conditions.

Into the twentieth century

Although the Victorians had introduced and pioneered then radical legislation, the prevailing society of extreme wealth existing alongside extreme poverty had changed little, but the First World War was to bring social change on an unprecedented scale. Throughout the world, societies were changing as the horrors of the war began to emerge. The revolution in Russia had overthrown its ruling class; Trade Unions in the UK were gaining considerable standing; men were returning to England physically injured and mentally disturbed; women had been employed in the previous male preserves and were not prepared to lose their emerging equality. There was a lot of pressure on government to act on ill-health issues such as tuberculosis and the still high rate of infant mortality. The emphasis was moving toward creating healthier housing and a better standard of living. In 1911, 9.1 per cent of the population lived at a density of more than two per room, and by 1921 this had risen to 9.9 per cent (Foskett 1999). The first rent controls had been introduced in 1915, and some vehemently argued that this was interference in the housing market and was responsible for the sector's decline, although wider changes were having a greater impact.

The government had promised 'Homes fit for Heroes' after the war and some 176,000 council houses were built under the Housing Act 1919 (Foskett 1999) to standards determined by the Tudor Walters Committee. This Act also provided a new duty for local authorities to survey the housing in their district. The house building programmes had several objectives, including the recruitment of post-war male unemployed workers and professional classes into house building to help quell growing domestic social unrest, as well as to respond to the very real need for decent and affordable housing. There were criticisms of some of these new programmes, including allegations of financial mismanagement in some areas, but, in general, large numbers of new homes were constructed to a good standard and, in general, house-building rates mirrored the Exchequer subsidy available. The Housing Acts 1923 and 1924 sought further to stimulate house building through new subsidy and, by 1927, more than 270,000 dwellings were constructed annually (Foskett 1999). The Labour Minister of Health, Wheatley, instigated a local authority house-building subsidy with trade union support. The majority of existing legislation concerning housing developments and general conditions was consolidated into the Housing Act 1925, by which time building for sale rather than for rent had become more profitable.

The interwar years saw a re-emergence of poverty with the decline in industry, a General Strike and the Wall Street Crash leading to the Depression in the 1930s. Public health worsened correspondingly, particularly in industrial areas. The problem of poor housing became so acute that the government was forced to act, introducing the Housing Act 1930 to instigate slum clearance and area improvement programmes, with accommodation provided for those to be re-housed. The Act gave local authorities powers to demolish or repair unfit dwellings leaving land for new development. In 1933, the Health Minister Sir Hilton Young promised £95 million for slum clearance to generate employment as well as to tackle poor housing. The 1930 Act also introduced new powers for overcrowding, which were built upon in the Housing Act 1935 by introducing a definition, guidance and powers to deal with overcrowding. It was envisaged that mass building programmes would both help reverse unemployment figures and replace slums with new local authority housing. Because conditions were so bad, it was relatively easy to equate improved health with improved living accommodation and the Housing Act 1936 added impetus to the recent legislation.

In the aftermath of the Second World War, housing took its place alongside the new National Health Service (NHS) as part of the Welfare State and was increasingly important politically. In 1944, the Dudley Committee recommended further standards for council housing. New towns were constructed in green-field areas, with many re-housed in new areas some distance from their previous lives in cities. The government pledged itself to regional regeneration, particularly since some half a million houses had been demolished by bombing raids. The 1930s' slum clearance programmes had been put on hold because of the war and about 173,500 houses included in slum clearance schemes had been disbanded. Additionally, unfitness had risen because routine maintenance had not been carried out. People's expectations had changed, and households were growing in number and changing in structure.

The first private sector housing grants were introduced under the Housing Act 1949, but with little overall impact on conditions. This was followed by the Housing Act 1954, which required local authorities to survey local housing – this revealed massive levels of unfitness. In Birmingham, 16 per cent of houses were unfit, in Manchester 33 per cent and in Liverpool 43 per cent (Foskett 1999). Post-war clearance programmes started again in 1954, but existing housing conditions tended to lag behind the general improvements in the standard of living envisaged through the Welfare State. The Joseph Rowntree Memorial Trust estimated that in the early 1960s, 29 per cent of households still had no bath, 28 per cent no hot water supply, 6 per cent no flushing lavatory and 10 per cent had to share a toilet. Immigrant communities were also growing but they faced much discrimination in housing and social opportunity. Many faced overcrowding and poor living conditions, particularly in inner city areas.

Housing law took a new direction and the Housing Act 1957 introduced the statutory standard of fitness and directed action to take in respect of unfit houses. Based largely on standards around 1919, it still did not include internal amenities (bath, hand basin, toilet), but housing had a new legal basis on which to move forward. Discretionary grants for rehabilitation and conversion were introduced in 1959.

Much happened during the 1960s that had an impact on attitudes toward poor housing and homelessness. The Parker Morris design standard, covering matters such as heating, floor and storage space, was introduced in 1961 and it remained mandatory until 1981. The government established the Milner Holland Committee in 1963 particularly to investigate private rented accommodation in Greater London, which drew attention to poor conditions in this sector. The first national House Condition Survey followed in 1967. The Rachman legacy and film *Cathy Come Home* in this period had changed many public perceptions of housing, homelessness and poverty. Housing pressure groups such as the Public Health Inspectors London Action Group (PHILAG), but more notably Shelter, were gaining momentum. The combined effect gathered support for an increase in local authority house building.

Meanwhile, local authority house-building programmes were shifting in emphasis from quality to quantity with the introduction of high-rise rapid build, concrete developments taking place on a massive scale during the 1960s and 1970s. Poor private housing was replaced with a new form of largely high-rise municipal estate, considered at the time to be an answer to the nation's housing problems.

Whilst there was, of course, a need for ending poor housing conditions, the policy response was not always beneficial. Many communities were broken up and have never recovered. Some clearance was unnecessary, and whilst the number of unfits may have reduced, this did not mean that the replacement municipalised housing was socially more desirable. Some thought that the housing crisis was over, and high-rises had answered the nation's housing problems, but new problems came to light, such as untested construction methods and inherent problems in design, as illustrated with the 1968 explosion at Ronan Point. Additionally, social problems were already starting to emerge with the new large-scale municipal schemes and emerging concentrations of welfare-dependent communities.

The Dennington Committee pointed to a new way to improve housing and its environment. It was also seen as more economical to provide some funding to poor housing before it required demolition and, in any event, most of the unfit housing had by now already been cleared, but much unsatisfactory housing remained, so a new power was introduced to tackle substantial disrepair. As a result, the Housing Act 1969 introduced General Improvement Areas in an attempt to target grants more effectively to areas of poorer housing and complement clearance programmes through area renewal. Even so, grant uptake was generally by better off people, and funding was not reaching the poorest sectors of private housing, in the owner-occupied or privately rented sector. This had resulted in gentrification and local people finding it harder to afford local housing. There were also allegations of 'winkling' whereby some landlords sought to displace tenants with a one-off financial payment following housing grant assistance and a subsequent rise in house-price (Balchin 1995).

The Housing Act 1974 introduced Repair Grants and sought to address how grant assistance was allocated, providing a more interventionist role and incorporating compulsory purchase. The Act introduced Housing Action Areas so that funds could be targeted into the poorest sector of private sector housing, preventing the need for clearance. Such legislation sought to use a mixture of public and private funds to arrest the decay in private sector stock so that it would not require early clearance. This was seen to be both more cost-effective and socially desirable. The problem of pepper-potted resources – grants being scattered across a local authority area – with some houses being improved and others not tended to undermine attempts at area improvement. As a result, enveloping schemes were introduced around 1979 so that local authorities could improve the external fabric of an entire block of houses, including windows roofs and walls, to encourage confidence in an area and owners to invest in internal works to their properties.

The Housing (Homeless Persons) Act 1977 for the first time acknowledged a growing issue of homelessness, and it provided a new duty for local authorities to meet the needs of the statutorily homeless. This remained so until the Housing Act 1996 controversially redefined homelessness and the 'right' to local authority housing accommodation.

Into the twenty-first century

Private sector renewal policies have favoured general rehabilitation over renewal of the private sector through grant assistance to owners to arrest housing decay. Despite massive public investment, housing conditions overall have remained relatively stable, although the age profile of the stock continues to rise (DETR 1998). Most local authorities have had to prioritise action in respect of poor housing and have had to make difficult decisions in respect of private sector stock. Some enforcement powers concerned with housing repair and HMOs, as well as all-renovation grants remain discretionary, and the climate of decreasing resources is likely to build up a greater backlog of houses requiring rehabilitation. In addition, there is little information available about grant applicants, other than by crude income determination and location, so it is more difficult to analyse trends to help ensure best targeting of available resources. Private sector housing renewal policy finds it difficult to address wider issues of disadvantage, which can continually undermine on-going renewal efforts, a key issue in policy sustainability.

The 1980s saw the rapid emergence of 'New Right' policies, with a withdrawal of the Welfare State, a growth in unemployment and poverty and a shift from the housing budget to the social security budget. In line with the ideology of deregulation, the Parker Morris standard was disbanded in 1981, allowing each local authority to decide its own housing standards. A form of 'privatisation' of social housing by schemes such as Large Scale Voluntary Transfer (from local government to housing associations) and Housing Action Trusts gained momentum. The Housing Act 1988 deregulated tenancies, and in addition temporary accommodation became increasingly important to house the homeless. The move from public to private sector housing was not without financial cost (e.g. Balchin 1995). There have been few long-term incentives for the private sector to invest in affordable renting in recent years, and there are no major forthcoming proposals to do so.

The ideology of the New Right from 1979 was to have a radical and substantial impact on housing policy. The Housing Act 1980 introduced the Right to Buy for local authority tenants, and the Housing Act 1985 consolidated legislation before the Housing Act 1988 introduced powers for new contractors to take over the local authority housing function with a range of new powers not available to local authorities. Grants were seen as increasingly ineffective, and a review of policy with wide consultation from 1985 to 1987 led to a fundamental change to the grants system and the development of home improvement agencies.

The resulting Local Government and Housing Act 1989 was designed to simplify grant procedures, linking means-tested mandatory grants to a new statutory standard of fitness (under the Housing Act 1985 as amended). It also introduced the concept of renewal areas to consolidating and adding to the best of previous area based schemes, although relatively few have been declared. Group repair schemes were introduced to encourage enveloping, grant assistance to improve the exterior of properties with a assumption that owners would be encouraged to invest in internal works, through a renewal of confidence in the area. Examples of selective clearance, rehabilitation and new-build arising from local authority housing grant and enforcement activity are illustrated in Figures 2.2 and 2.3.

The mandatory link between fitness and grant was broken by the Housing Grants, Construction and Regeneration Act 1996, largely due to acute funding problems, leaving local authorities to use their discretion in delivering local private sector housing renewal. There are current proposals to finance public and private sector housing from one budget, combining private sector renewal grants with the main capital budget. There is a risk that the private sector will lose out on funding as local authorities use available resources to regenerate their own housing stock.

Figure 2.2 Selective clearance. This illustrates how selective small-scale clearance can remove redundant housing stock and create space for a new street layout

Figure 2.3 Infill development. This illustrates how new build, following selective clearance, can complement an existing area without too much community disruption. The new build properties are set back from the road to allow for parking, a general improvement to accommodate changing needs in older terraced housing stock

Moving forward – housing and sustainability

Gro Harlem Brundtland, former Chair of the World Commission on Environment and Development, defined sustainability in 1987 as follows: 'Humanity has the ability to make development sustainable – to ensure that it meets the needs of the present generation without compromising the ability of future generations to meet their own needs' (cited in Sustainability 2000). The ability to sustain ourselves, our housing and our communities requires some basic changes and local authorities can move in the right direction through the way in which they deliver their housing functions. Cyberus (1999) suggests that activities are sustainable when they:

- use materials in continuous cycles;
- use continuously reliable sources of energy; and
- come mainly from the qualities of being human (including creativity, communication, coordination, appreciation, and spiritual and intellectual development).

Conversely, activities are not sustainable when they:

- require continual inputs of non-renewable resources;
- use renewable resources faster than rates of renewal;
- cause cumulative environmental degradation and extinction of species; and
- undermine others' well-being.

It therefore follows that sustainable development would be based on the four principles of sustainability: futurity, environment, equity and participation. In terms of housing, there are wide-ranging implications.

Sustainable housing requires a community based and multi-agency approach to be successful and, in effect, to be self-regulating in the longer term. It requires input from all those involved in housing, planning, the environment and community services at local level. Some local authorities have maximised the potential of Local Agenda 21 as a framework to deliver sustainable and locally developed housing policies (LGMB 1995). This process requires that local authorities develop sustainable indicators, many of which can be drawn from lessons from past private sector renewal policies – what has proven sustainable in the longer term and, conversely, what has not. Forrester (1998) argues that the planning function is key as the number of households continues to grow, and the demand for both brown- and green-field sites requires close control. Construction and design processes are also key, so that new house building and rehabilitation of existing housing both uses renewable resources where possible, and that maximum energy is conserved during the property's lifetime.

Sustainability also incorporates lifestyle issues. House design needs to cater for an ageing population and changing households needs; to be located to minimise travel; to incorporate renewable materials and conserve energy and to be located sensibly to promote health and housing issues; and to have an optimum density with a good infrastructure (Forrester 1998, Conway 2000).

Such issues pose a huge challenge to a local authority housing function, particularly in the private sector, where much work is a response to existing housing conditions, rather than the development of new. There is however an increased potential to incorporate sustainability issues into some areas of housing, which are covered throughout this book. Renewal action will almost always lead to some improvement in the lifetime of the property; the Home Energy Conservation Acts 1995 and 1996 require local authorities to reduce carbon dioxide emissions; renewal areas can provide the impetus for wider socio-economic change; and so on.

The government's recent document *Regeneration that Lasts* (DETR

2000) refers to elements of sustainability in social housing rehabilitation and notes the importance of well-thought-through exit strategy that happens at the right time to enable project sustainability. This strategy, which also has implications for private sector renewal areas, includes the on-going ability to:

- preserve the results of capital works, including environmental works;
- sustain improvements in housing management; and
- sustain the 'Housing Plus' agenda.

In terms of private sector renewal, the equivalent would be for the local authority to exit a renewal area at the point at which the community was displaying the ability to sustain itself locally through new community initiatives, employment and local services, with a marked improvement in housing conditions and infrastructure.

Summary

- Edwin Chadwick first linked housing conditions and poor health around 150 years ago.
- Legislation addressing poor housing conditions received much government support after the First World War.
- Housing policy has followed political ideology and changed markedly since the Second World War.
- Recent years have seen a gradual withdrawal of the welfare state, moves away from municipal housing stock, a targeted grants system encouraging area action with public–private partnerships, and moves toward sustainability through multi-agency approaches.

2.2 Housing and health

Outline

Housing and health is a wide-ranging subject covering all issues of physical and mental health that might arise from poor living accommodation, or conversely well-being that might be promoted by decent accommodation. This section reviews issues of housing and health and introduces the potential for positive action in dealing with private sector housing.

General background

The World Health Organisation (WHO) defines health as being 'a state of complete physical, mental and social well-being, and not merely the absence of disease or infirmity', an optimistic and all embracing statement (WHO 2000). A healthy body needs to be in a state of equilibrium and the Greek term 'homeostasis' means 'staying the same', so that it can cope with changes to its external environment. Illness, then, is the body's response to harmful environmental forces such as poor housing and a poor local environment. A person's housing conditions plays a crucial role in their physical and mental well-being; the better the housing, the better the health. Conversely, the worse the housing, the worse the health is likely to be.

The Greeks first established a link between housing and health, but it was not until the early sanitary reformers such as Chadwick made a link in the UK, when the term 'slum' was introduced to describe poor housing conditions. The Victorians began to keep statistics that illustrated a profound link between premature death and impoverished conditions, but linked poor housing and ill-health with smells. They, however, took a broad view of health – including relating overcrowding to potential harm to mental and moral health by inadequate separation of the sexes, but their responses nevertheless began to reduce housing-related diseases such as tuberculosis.

Linking housing and health

Whilst the link between poor housing and health continues to be recognised it is difficult directly to link by empirical evidence. It is generally described in terms of negatives rather than in terms of good housing promoting well-being. This is because there are many other factors that affect ill-health, such as social disadvantage, poverty, inadequate diet, poor working conditions or unemployment, lack of medical care, and so on. Attempting to measure the health impact of poor housing is difficult, particularly in cases of mental health. There are few empirical studies available, and a general problem in coordinating health and housing information between various organisations such as architects, housing officers, doctors, EHOs, social workers, policy developers, and so on directly relating to poor conditions. However, it is generally regarded that the combination of factors that make up unhealthy housing has an effect on health (Audit Commission 1991, Ransom 1991, Townsend *et al.* 1992).

The English House Condition Survey (DETR 1998) continues to point to the fact that though most of the worst condition housing is in the owner-occupied sector (because this is the predominant tenure); as a percentage, the private rented sector suffers the worst housing, particularly to the most disadvantaged groups, such as the homeless in temporary accommodation and those otherwise unable to secure decent accommodation in the housing market. The Audit Commission Report on Healthy

Housing (Audit Commission 1991) only comments on physical aspects on ill-health – and makes no mention of mental ill-health such as depression. The government tends to focus on a medical rather than social model of health. Measurement of health is usually based around occupational class, which excludes a lot of groups, such as people with a disability, who have special housing needs. There are also different class perceptions of health and how articulate people are in visiting their doctor and so on. Such a focus tends to be individual and looks at cures for health, not social and environmental factors that may prevent ill-health and promote well-being. Dependence on mass social housing is politically no longer seen to be the answer as it was in the heyday of the Welfare State and the health debate in terms of private sector housing is invariably reactive. Some links between health and housing are now considered.

Homelessness and temporary accommodation

Many current issues on housing and health focus on an increase in the use of temporary accommodation, which aggravates many of the issues discussed below. Some local authorities make extensive use of accommodation rented from private landlords, mostly to offer temporary housing to those accepted as unintentionally homeless and in priority need (see the definitions in Section 2.5), for whom there was no suitable accommodation available in the social rented sector. Temporary accommodation can be provided either direct by private landlords to nominated households, or through an intermediary such as a housing association as leaseholder or manager. Temporary accommodation comprises many types of accommodation, normally financed through housing benefit, which the DETR (1997) categorises as follows:

- Private sector leasing.
- Housing association leasing.
- Housing association as managing agents.
- Assured shorthold tenancies.
- Discharge of duty.
- Bed and breakfast.

Of the above categories, bed and breakfast accommodation is frequently in poor condition, and dependence on this sector for temporary accommodation has fallen in recent years in favour of other types. However, a substantial number of households are still housed in bed and breakfast establishments, and it is this sector that is mainly considered here to illustrate health and housing issues.

The London Research Centre (LRC) established the Bed and Breakfast

Information Exchange (BABIE) in 1988 as a response to poor conditions in hotels being used by London boroughs for temporary accommodation. At this time there were about 20,000 households in temporary accommodation, some 7,000 of which were placed in bed and breakfast accommodation, although the figure has now dropped to around 3,000 (LRC 2000). Owing to increasing pressure on accommodation, as well as increasing costs, many were housed outside of their own boroughs, so housing conditions became another local authority's responsibility. BABIE was formed to coordinate action and agree recommended prices for rooms across local authorities and it has prepared a common grading system for hotels, which EHOs inspect. The details are of location and condition, including amenities, etc., and are centrally collated by BABIE. This has led to a gradual improvement in such temporary accommodation, and disuse of accommodation that remains unsatisfactory. However, there is still some way to go.

The effect on health from living in bed and breakfast-type temporary accommodation has been well documented. Day-to-day living in overcrowded conditions and sharing insufficient facilities with strangers has some impact on ill-health. Such lifestyles frequently result in accidents, small babies, less immunisation of children, poor nutrition and depression arising from uncertainty and poor conditions (Conway 1988). Providing comprehensive health services to temporary accommodation residents is difficult; because it is temporary, many find it difficult to register with a GP and so do not; some GPs do not wish to take them on for financial reasons. They may end up as no one's responsibility in the hospital system. Closure of many specialist hospitals has left many people living in the community with inadequate support in temporary accommodation.

Many fall outside of the statutorily homeless definition and self-place themselves in temporary accommodation, which can aggravate their health, housing and social needs. They may have less support, but similar health and support needs, to those deemed statutorily homeless. Many asylum seekers fall into this category (see Section 2.6), and may already be disadvantaged because of accessing low-cost unsatisfactory private rented accommodation with inadequate local facilities, often compounded by language and cultural differences, a general lack of resources and simply having nowhere else to go. Here, housing need and conditions can be at their most acute and multiple disadvantage needs to be addressed.

Being accepted as homeless and in 'defined' temporary accommodation or otherwise housed in the private rented sector is only one side of the story. The deregulated assured shorthold tenancies introduced by the Housing Act 1988 place considerable pressures on private sector tenants and their families. There is without doubt some impact on tenants' emotional well-being, and the constant pressure of a possible end of tenancy and having to move after 6 months is, at the very least, stressful.

Rooflessness

Homeless is not the same as roofless. The number of people sleeping rough is unknown and it is difficult accurately to quantify those with nowhere at all to live, but the figure may be up to 96,000. Many are discharged mental health patients, some of whom struggle to cope in 'formal' housing. It is difficult for GPs to take on, and keep track of, homeless people who may go to a specialist clinic, but follow-up is difficult unless the patient self-presents. Crisis, the national charity supporting single homeless people with no legal rights to accommodation, seeks to help individuals to rebuild their lives and to move into sustainable housing. Crisis's Health Action for Homeless People initiative seeks to improve homeless people's access to a range of quality health and social care services (Crisis 2000). It estimates that 65 per cent of premature deaths were probably preventable given proper housing and good health care. The sector displays high, and growing, levels of tuberculosis.

Cold and damp

Cold and damp is perhaps the most familiar aspect of health and housing to the majority of people. Cold and damp are intrinsically linked with poverty. Numbers in fuel poverty (see the definition in Section 2.7) rose from 5.5 million in 1981 to 7 million in 1991, particularly among people at home all day who require more heating. This is clearly more expensive, so cheaper methods may be used, which aggravate damp, as does drying clothes indoors because there is nowhere else to do so. The poorest 20 per cent of households spend 12 per cent of their budget on fuel, whereas the wealthiest 20 per cent spend 4 per cent. VAT on fuel has aggravated this fuel poverty (Boardman 1991).

Cold and damp can be closely related to construction type. It is relatively more expensive to heat poor older housing, particularly in the private rented sector where landlords have little legal or financial incentive to invest in energy efficiency. Older people in rented accommodation those in lower occupational groups are less likely to have central heating and there is less dampness and condensation in centrally heated accommodation. It can also lead to decay of building fabric. Ironically, there has been a rise in complaints of dampness and condensation through improvements being carried out which have reduced ventilation levels, such as by sealing chimneybreasts or installing double-glazing.

Ill-health effects include increased levels of hypothermia, physiological changes in the body, heart attack, stroke, cardiovascular and respiratory disease (especially in children); asthma and mould sensitivity, and stress and depression from visual effects of mould growth (Ormandy and Burridge 1988, Boardman 1991, Lowry 1991, Arblaster and Hawtin 1993, Ineichen 1993, Markus 1993, DETR 1999, DoE 1996).

Legal standards for heating and insulation in existing dwellings are minimal, as discussed further in Sections 2.7 and 4.2. Normally all that can be required is a fixed heater in the main living room and provision for heating (sockets) in other rooms, with minimum loft insulation. EHOs regularly find that even this requirement is not met in many premises. Figure 2.4 illustrates the only heating source in a privately rented premises – clearly insufficient to provide adequate heat and extremely expensive to operate. The condensation dampness shown in Figure 2.5 illustrates the condensation mould growth that was a direct result of inadequate heating, ventilation and loft insulation in this instance. Possible legal remedies for such a situation are explored in Chapter 4.

Noise pollution

Noise pollution is closely related to construction. Temporary accommodation can aggravate noise nuisance due to overcrowding and poor noise

Figure 2.4 Inadequate heating source. This electric bar heating source was the only heater in the dwelling – it was incapable of providing adequate background heat for the room and was expensive to operate. Its short-term use would be likely to aggravate temperature differentials and cause condensation

Figure 2.5 Condensation and mould growth. The condensation and mould growth result from the poor heating supply in the living room as well as inadequate insulation and ventilation to this room (the bathroom). The mould growth clearly indicates how it is less likely to proliferate where insulation is provided, and there is no mould growth where the roof joists provide some insulation. In such cases, requiring the landlord to provide an adequate fixed heating source in the main living room and loft insulation, combined with advice to the tenants, would help resolve the situation

attenuation due to inadequate building materials, poor design and insulation. Complaints relating to noise have risen about twenty-fold in the past 20 years (Ineichen 1993). Tension from noise, such as loud music or regular arguments, can cause major problems between neighbours and increased stress levels for sufferers. Remedial action falls to the EHO, often with police support, but enforcement is often extremely difficult.

Space standards

Statutory overcrowding is now uncommon except accommodation such as HMOs, where conditions are aggravated by a variety of other factors including sharing amenities and the means of escape from fire, which can lead to accidental injury as well as to increase the risk of fire. This is particularly so in bed and breakfast accommodation where a study has found almost 50 per cent to be statutorily overcrowded, lacking adequate facilities and providing little control over occupier's personal space, such as in communal areas (Conway 1988). The health effects of overcrowding are related to an increased incidence of infectious disease, both minor ailments and more serious, including tuberculosis, which is currently rising. Stomach cancer in adults correlates with overcrowding in childhood. Under-crowding can also be a problem, with loneliness, isolation and fear of going outside (Lowry 1991).

Domestic accidents

There are about 5,500 fatal accidents per year in British homes, and another 2.2 million non-fatal accidents requiring hospital treatment, another 900,000 requiring GP treatment which cost the NHS some £300 million annually. Domestic accidents are the commonest cause of death among children. Childhood accidents, including those leading to permanent disability, show a significant correlation with social class, and a further association with unemployment, overcrowding tenure, education, etc. Older people show an increased likelihood of accidents, such as from poor mobility and vision, and open fires rather than central heating. Internal layout and design such as awkward stairs, widely opening windows and so on can contribute to an increased accident rate.

Many features are introduced during domestic planning phases, but others can be incorporated into existing homes such as good lighting, fire guards, grab rails and stair gates to stairs, safety glass, and smoke detectors. Many of these features could be incorporated at relatively little cost at the design phase.

Accident levels are higher in temporary accommodation, which is ill-designed, ill-equipped and ill-maintained (Conway 1988, Lowry 1991, Arblaster 1993). These hazards combined with makeshift cooking and heating arrangements, overloaded electrical installations and inadequate means of escape are particularly pronounced in temporary accommodation where homeless families are regularly placed. HMO residents are ten times more likely to die in a fire than residents of other dwellings (Home Office 1989).

Increased reliance on artificial lighting due to changes in the Building Regulations might cause physical damage and increase the possibility of accidents. There is an increased risk of accidents (especially to older people) in poorly lit accommodation, which might be improved by better light bulbs and decoration to increase illumination.

Depression

Whilst depression is not exclusively caused by poor housing, there is no doubt that living in poor housing conditions can aggravate feelings of isolation and desperation, leading to the development and maintenance of mental ill health (Arblaster 1993, Ineichen 1993). The stress of day-today living in an unfamiliar area, overcrowded conditions and sharing facilities with strangers cannot be understated. Temporary accommodation is disrupting, uncertain and often means the loss of a social support network (Arblaster 1993). It is not hard to see how a mixture of poor construction and insulation, a lack of space, delays in necessary repairs, dampness, pest invasion and so on combined with wider factors such as crime, harassment and living in a run-down area with few services would effect mental health. Women at home, lacking social interaction, privacy and a leisure time due to child care responsibilities and in poor housing conditions are particularly likely to become depressed (Brown and Harris 1978).

Air quality and indoor pollutants

People spend 80 per cent of their time indoors, or far more if a vulnerable group, such as older people, the disabled, the young or unemployed. Ironically, air quality has decreased with energy efficiency measures. The effect of different pollutants varies with time as well as individual susceptibility. Carbon monoxide poisoning has attracted much publicity due to recent fatalities caused by faulty heating appliances. Radon exposure (in granite areas) is long-term and can lead to lung cancer; asbestos exposure is only a problem if it is friable or if renovation works are going on; formaldehyde can result from cavity wall insulation; nitrogen dioxide from burning fossil fuels and gas cookers. Legionnaires disease and domestic smoking also fall into this category.

The quality of indoor air can always be improved by increasing ventilation, removing or modifying the source of pollution, filtration or changing the occupant's behaviour. Ventilation helps to maintain or improve indoor environmental conditions so that heating and cooking appliances operate effectively, humidity levels are kept below critical levels associated with house dust mites ability to breed, and so on. Problems frequently occur where accommodation is shared and maintenance of equipment is the responsibility of another person, such as a landlord. Those living in flats, hostels, bedsits, and bed and breakfast accommodation are most vulnerable, often because dangerous appliances have been reconnected or where owner-occupiers cannot afford to maintain or replace faulty equipment.

The increase in asthma, particularly to children, has been heavily publicised in recent years. Asthma is a respiratory condition and symptoms include wheezing, coughing, shortness of breath and tightening of the chest. It is not possible to identify what is responsible for the increase in asthma, which seems to vary between individuals, and may be a result of antibiotics weakening the immune system, eating processed foods, lack of exercise, smoking during pregnancy, and so on, and the UK has amongst the highest rates of asthma in the world (BBC 2000). Whilst not the sole trigger for asthma, there is little doubt that symptoms are at least worsened, if not caused, by poor internal environments, where house dust mites and their waste products as well as condensation and associated mould spores can aggravate asthma.

Sanitation

Legally speaking, drinking water should be 'wholesome'. The European Union's Directive on drinking water quality came into force in 1985, when some domestic water failed the 100 per cent pass standard required. The water authority's responsibility is as far as the stopcock in the house, when it becomes the householder's responsibility. This is why pipework can be a source of contamination, such as lead pipes in soft water areas. Local authorities sometimes carry out checks on water supplies, especially private water supplies and home treatment plants where known about.

Waste disposal tends not to be an issue in the UK except where problems arise with individual pipework, cesspits and so on, where public health risks are obvious. The English House Condition Survey (DETR 1998) illustrated that some households still lack basic amenities such as an internal toilet, especially in the private rented sector. Many amenities have to be shared in HMOs, which can present problems of hygiene where management standards are low, possibly leading to diarrhoea, particularly to high-risk groups such as children.

Waste disposal

If allowed to build up, such as in communal areas where no one takes overall responsibility, domestic waste can lead to secondary problems of vermin. Rat and cockroach populations are increasing and can be vectors of disease, especially where design factors encourage their multiplication. Refuse can also present problems of accidents, fire and stress.

High-rise flats and municipal design

As a mass, low-cost response to housing need in the 1960s, many high-rise flats, though providing a decent internal living space with internal amenities, were poorly designed, developed and constructed to provide an unsatisfactory outside environment. Many had inherent problems of the low structural and insulation properties of concrete and metal supports, with asbestos insulation. Poor thermal qualities and cold bridging from supporting beams is linked to high heating costs, compounded by cheaply installed heating that is inefficient to run. Besides inherent design faults, they can be difficult to manage, leading to problems with refuse chutes, lifts, vandalism, lack of operable lighting and crime (Coleman 1990). Many of these issues would be difficult to address because they are 'designed in'. There are few places for children to play outside because of safety. Again there is the issue of who is responsible for communal areas.

Health problems also include feelings of isolation; lack of choice and control in housing allocation with a loss of family and friends network; lack of community, facilities and shops; mental ill-health has been found to increase with the number of floors and women are particularly affected. Poor construction, use of materials and design also lead to damp housing. The Child Accident Prevention Trust suggested that architectural design would substantially reduce the number of child hospital cases (currently 250,000 annually). Social exclusion has become an issue in many such estates, although local authority (and ex-local authority) estates are by no means the sole location of exclusion.

Special needs housing

Housing for older people should not be thought of *en masse*, but for individuals who may have special needs increasing with their increasing vulnerability, such as maintaining body temperature, poor eyesight, restricted mobility and deteriorating memory in some cases. 'Lifetime Housing', incorporating, for example, wider doors, designing kitchen for later adaptation, installing lever taps, is finding it hard to enter design processes. Older people may have increased heating requirements due to reduced mobility, but some struggle to afford this. Severe weather payments are usually too little too late and insulation grants are inadequate. Isolation may delay help arriving in case of accidents. People may not wish to admit they are elderly and their family may pressurise them into decisions about their housing, or having an alarm system to call for help. Violent crime against older people has increased in recent years. With an increasingly elderly population, there is more choice than ever for housing possibilities, but people's needs are continually changing and the resultant financial implications can be considerable. Often older people are forced to move suddenly due to illness. Many older people are well off and in good health and some suggest that they are given unfair advantage over more needy groups.

Housing and disability

There is little housing choice available to people with learning and physical disabilities, particularly for young people. Many live in the community relying on carers for support, but they may wish to have homes designed to live more independently. People with disabilities do not wish to be labelled or stigmatised, but often, for economic reasons, small estates have been developed to cater exclusively for them, which can be isolated from the wider community. There is tremendous inequality between those with disabilities, because there is so little purpose-built accommodation. Accommodation often requires adaptation to meet specific needs, rather than some simple features being designed-in at planning stages (see Section 4.7).

Housing and health – breaking the link

Ormandy and Burridge (1988) suggest that the 1939 American Public Health Association's (PHA) 'Basic Principles of Healthful Housing' is still a useful background to housing assessment. It lists four fundamental categories by which housing standards can be measured: physiological needs, psychological needs, protection against contagion, and protection against accidents. The PHA, even in 1939, recognised the importance of not just individual housing conditions and their relationship to health, but the wider issues of housing within its community. The PHA considered that the local environmental quality, noise levels, space for exercise, provision for a normal family and community life, and so on were equally important as, for example, adequate heating or a safe water supply.

The World Health Organisation (cited in Ransom 1991) described the complexities of healthful housing both within the house itself and its local environment. It reinforced the view that housing is not just about the avoidance of illness but it provides a living environment for betterment of health. It set targets relating to health and housing to be achieved by 2000 within available resources. Unfortunately, poor housing, particularly to lower-income groups in the private sector, remains at unacceptably high levels.

There are two main issues as to why this is so. First is that housing renewal policy is based within legislation on already defined standards enforced by EHOs (Ormandy and Burridge 1993). There is no real scope to embrace all health professionals, GPs, community workers, health visitors and so on to develop the promotion of healthy housing. Second, and of increasing importance, is the use of temporary and other unsatisfactory accommodation in the private rented sector to house the homeless and those unable to secure affordable accommodation elsewhere.

EHOs are the key professionals involved in delivering legislation to promote housing conditions in terms of housing fitness, nuisance, pest control and public health in the private sector through a range of measures described in Chapter 4. There is also scope for campaigning for improvements on housing and community. Much poor public sector housing is currently being renovated or redeveloped, with a move away from tower blocks to more traditional designs. In improving housing and health, it is important to target resources to the worst housing identified, which will inevitably include HMOs and the bottom end of the private rented sector as well as housing meeting the specific needs of the occupiers. There is also a clear need for increased liaison between health professionals and multi-agency working. Particularly relevant to private sector housing, the government is currently considering a housing fitness rating scheme that seeks directly to relate health and housing issues, and to introduce a national licencing scheme for HMOs. These are discussed further in Chapter 4.

Also encouraging are moves toward a reworking of public health issues since the late 1980s (Ashton and Seymore 1988). This revision of pubic health – the New Public Health – seeks to draw together partnerships (see Section 5.6) of those involved to develop new and innovative ways forward in promoting well-being. This represents a substantial move away from traditional static organisations delivering predefined and segregated services. The new public health combines health promotion, healthy alliances, issues of equality and empowerment through new styles of management and service delivery in a corporate approach. The success of such approaches can be seen in the many examples of well-attended Healthy Living Centres, which offer a new approach to, and interest in, community health in its holistic context.

The recent White Paper, Saving Lives: Our Healthier Nation (DoH 1999) has sought to consolidate and develop such initiatives in order to tackle the complex causes of ill-health (personal, social, economic and environmental). Partnerships and joined up government action are seen as key in order to extend and encourage healthier lives by tackling and reducing health inequalities. Housing quality is recognised in impacting health across all tenures, and the government's current policy initiatives to address poor private sector housing include extending Home Improvement Agencies and the proposed housing health and safety rating system, as discussed further in Chapter 4.

Summary

- Health and housing issues are wide ranging and can be aggravated or resolved by a variety of issues.
- People with special needs, such as the elderly and those with disabilities, sometimes require particular adaptations to enable them to make use of their existing accommodation.
- Some health and housing issues can be resolved relatively easily and inexpensively, but others are far more difficult to address and can be reinforced by a complex matrix of inequality.
- Existing legislation tends to be reactive and cannot address wider issues of disadvantage.

- New legislation is set to focus more directly on the health and housing relationship rather than concentrating on physical standards as at present.

2.3 Role of the private rented sector

Outline

A definition of the private rented sector was included in the Introduction. Here the private rented sector is defined as that which is let as a business enterprise to make a profit, so does not include housing associations or other not-for-profit lettings. This section briefly overviews the role and function of the private rented sector, its rise and fall over the course of this century, and the way in which government policy is encouraging its current role to provide, in essence, a new privatised form of social housing heavily subsidised by housing benefit.

Any mention of the role and function of the private rented sector can raise all sorts of value and ideologically laden debate when considering housing provision. At the turn of the twentieth century, this sector provided 90 per cent of the nation's housing. By the end of the century, that amount had fallen to less than 10 per cent. This section looks at the reasons for the decline of the private rented sector and the ideological arguments that seek to provide these reasons. It explores rent levels and client groups, conditions in the sector and the current 'privatisation' of the local authority role in meeting housing need, which is discussed further in Sections 2.2 (Homeless and temporary accommodation) and 2.4. Although some housing authorities are reliant on the private rented sector to meet local need, particularly across London, there is a surplus of hard-to-let local authority accommodation in other areas. There are, therefore, differences in supply and demand across the country.

Fall and rise of the private rented sector

Interpretation of the decline of the private rented sector falls largely into two camps. First is the view that it is not allowed to operate as a market. This view is mainly concerned with what is perceived as over regulation, which prevents the housing market from operating effectively. The second view is that the decline is due instead to wider social changes, fiscal policy favouring owner occupation as a tenure and the provision of low-cost social housing by a different provider.

The reality is extremely complex and tied up with wider social issues. Could the decline have been due to over regulation? The argument seems simplistic because even before the first rent controls in 1915, the private rented sector had become a less attractive investment to landlords. Capitalists wanted cheap labour, but such labour could not afford market rents. Labour power was relatively strong and women were taking a greater role in the workplace. The resulting rent strikes led to the first controls of the sector. The private rented sector was already becoming unprofitable and the tenure shift was already underway because building societies started to lend more widely. The fiscal advantages of owner occupation were already becoming clear.

Other factors were also affecting the private rented sector. Slum clearance from 1930s, but especially from the 1960s, and consequent municipalisation of stock, particularly between 1961 and 1975, provided an alternative to private renting. Owner-occupation was becoming the preferred tenure. By the late 1950s and 1960s, the Rachman legacy had created image problems of the sector, mainly due to extortion and deliberate avoidance of regulatory structures.

The private rented sector – a renewed purpose

The Conservative government elected in 1979 sought to rejuvenate the private rented sector, arguing that the decline was largely due to over regulation so that renting was no longer an attractive investment option. It, therefore, sought to deregulate security of tenure and enable market rents to be charged through the introduction of assured shorthold tenancies under the Housing Act 1988. There was some increase in the sector after this, but this was to a large extent due to wider economic factors, including a slump in home ownership, a declining social housing subsidy since the mid-1970s, an inability to obtain a mortgage due to 'flexible' work, short-term contracts, and so on. Considerable government funds were pushed into the private rented sector during the late 1980s in an attempt to rejuvenate it. For example, the government tried to appeal to institutional investors through the Business Expansion Scheme, but this largely failed, as did Housing Investment Trusts. Cross-tenure initiatives such as Housing Associations as Managing Agents proved expensive. What became apparent was that the days of institutional investment in the private rented sector were well and truly over, and that most of the sector's growth was from 'sideline' landlords.

Current landlords are a disparate group. Recent research (DETR 1996) found that landlords were generally more positive about letting because they could charge market rents, find gaining possession easier and believed the sector generally to have a better image. Most lettings are by landlords as a sideline activity; most are owned as private lettings and few proportionally by private lettings companies. It tends to be a small-scale activity, and one-quarter of landlords have only one letting. Around one in ten landlords rent as they cannot sell because of the nature of the housing market and their portfolios are likely to remain similar in the foreseeable future.

The private rented sector currently comprises some 2 million properties, making up approximately 10 per cent of national housing stock. The private rented sector is seen largely in terms of investment potential to the landlord rather than housing provision, at low rent, to the tenant. But the private rented sector is an imperfect market; a substantial sector of housing for rent is provided by the state (in various forms) and housing benefit plays an increasing role in funding the sector. There are some issues concerned with renting housing as an investment. First, why would a landlord choose to rent a property rather than take an alternative investment opportunity? And, in addition, why would a low-income tenant choose to rent privately?

Put simply, the private rented sector operates as a form of 'dual market'. It caters well for those able to afford market rents and who perhaps require more flexibility in housing than owner occupation can offer. However, there is another side to this market that caters for a very different client group and offers a very different standard of accommodation for those unable to afford accommodation elsewhere; those who have little, if any, choice in their housing. This sector is heavily subsidised by housing benefit, but it proves difficult to regulate in terms of security of tenure or condition of property. This, of course, is the sector where environmental health enforcement activity largely falls, where officers regularly find a totally unregulated private rented sector through evasion or omission, deliberate or otherwise.

Establishing a suitable rent level

Rent level is broadly based on capital value, the value of capital (for sale) rather than of revenue (for rent). A regularly asked question is why someone would choose to rent out a property rather than invest their money in another way. This question is difficult, perhaps impossible, to answer, because landlords are a heterogeneous group and in the property market for different reasons. Much debate concerning the private rented sector focuses around the adequacy or otherwise of rent level to the landlord. Little of the literature is concerned with the capital asset and investment potential of the property itself, but on revenue issues and the maximum return on the landlord's investment.

Policy during the 1980s was driven toward deregulation of tenancy legislation arguing that this would encourage more landlords into the market, thus unleashing a supply of accommodation for private renting. The emphasis focuses greatly on a shift from public to private housing in meeting the needs of the homeless. And what this of course required was someone to pay these new deregulated rents.

Herein lies the problem. It has already been seen that those in a position to buy generally do, because this is generally more secure as well as offering financial advantage. Those able to access social housing secure a tenancy let at low rent, and normally on a secure tenancy. This leaves many low-income groups who have few, if any, choices in the housing market, with nowhere to turn but the private rented sector. As Cowan (1999: 305) observes, 'the sector's primary client base in recent years has been the provision of temporary accommodation to the young and/or mobile who have difficulties in accessing accommodation in the social sector'.

In terms of housing provision at an 'affordable' rent, there is a fundamental dilemma at the core of the private rented sector. Landlords are only likely to enter the market if they can expect a reasonable revenue return, so market rents need to be charged. For those able to afford such rents, home ownership tends to be the preferable tenure. The state subsidises those unable to afford those rents via housing benefit, but is this really a subsidy to the tenant or a subsidy to the landlord? Housing benefit comes with no strings attached – housing conditions are not checked or qualified before it is paid.

The private rented sector and housing benefit

Housing benefit is paid across rented tenures, with 60 per cent of claimants living in council housing, 19 per cent in registered social landlord properties and 22 per cent in the private rented sector, and there are geographical differences to housing benefit distribution.

Housing benefit has increased substantially in recent years, shifting spending from direct bricks and mortar subsidy to personal subsidy. In 1979, the balance was 84 per cent bricks and mortar subsidy to 16 per cent personal subsidy, and by 1998/9 this had altered to 27 per cent bricks and mortar subsidy to 73 per cent personal subsidy. In 1978/79, £2.3 billion was spent on housing benefit (at 1998/9 prices) (DETR 2000). Prior to the Housing Act 1988 and the new 'deregulated' tenancies, the housing benefit bill was £3.9 billion and by 1995/6 was £11.9 billion (Cowan 1999). By 1998/9, the figure was £11.1 billion and is forecast to rise at 1.4 per cent per year in real terms until 2001/2, both due to changes in rent levels and the number of people getting it (DETR 2000).

Fifty per cent of private rented sector tenants receive housing benefit, showing the considerable demand for this sector from low-income households. Forty per cent of landlords do not want to rent to housing benefit claimants, but 50 per cent need it (Cowan 1999). Many landlords see reliance on housing benefit as a constraint and benefit tenants are frequently viewed with suspicion. Perceived reasons for this include delays in processing housing benefit, uncertainties in paying deposits, uncertainty about benefit level and attitudes toward personnel involved in delivering benefit. How does government reliance on the private rented sector fit in with this? Does the private rented sector have the capacity and willingness to provide?

The Conservative government response to the rising housing benefit level, largely created by their own policies, was to try to cut expenditure on housing benefit, both directly and indirectly. What was formerly achieved through rent control was achieved through cuts to housing benefit as landlords sought to raise the lid on the market (Cowan 1999). The current Labour government seems committed to maintain the supply side of the private rented sector through continuing with the tenancy changes favouring landlords since the Housing Act 1988 (DETR 1996). The current Green Paper (DETR 2000) suggests a reform of the housing benefit system rather than a return to a bricks and mortar housing subsidy and changes are proposed to:

- improve customer service;
- reduce fraud and error;
- improve work incentives (through reducing barriers to work); and
- explore other options to support housing policy.

Housing conditions in the private rented sector

Originally built to house the working classes, this part of the private rented sector was built to lower standard than for other rental groups from whom higher rents could be expected. The majority of the private rented sector is pre-1919 and this is one of the main reasons for its generally poor condition in relation to other tenures. The English House Condition Survey (DETR 1998) shows that the greatest proportion of unfit properties is in the private rented sector, and those low-income groups, particularly ethnic minorities, are represented in this sector. This is because ethnic minorities are frequently excluded from other housing options by a matrix of discrimination elsewhere. Low-income groups, on short-term tenancies, are unlikely to effect repairs or improvements themselves, and may fear asking their landlord to carry out essential works for fear of eviction.

Ironies of policy

There is current cross-party consensus on an increased role for the private rented sector in responding to increasing homeless and decreasing social housing. Seeking to open up the private rented sector is just one in a series of policies that seeks to encourage 'choice' and 'diversity' in provision. In delivering their housing responsibilities, local authorities have a duty to consider 'other suitable available occupation' and to provide 'such advice and assistance as the authority considers necessary', which is of course down to interpretation.

The fact remains that there is simply inadequate social housing to meet the needs of those accepted as homeless, let alone others unable to afford and secure decent housing. Local authorities have to find accommodation somewhere, and that somewhere currently seems to be through tapping into the existing and potential supply of the private rented sector. The costs of providing private rented housing as a form of temporary accommodation (see Section 2.2, Homelessness and temporary accommodation), or otherwise subsidising it through housing benefit, have risen markedly sine the Housing Act 1988. Balchin (1995) saw such economics as distorting demand, rather than facilitating an adequate supply of decent social housing for rent.

The private rented sector and the Green Paper

In England, one in ten people live in the private rented sector, far lower than other developed countries (DETR 2000). The current Labour government is committed to a healthy private rented sector for flexibility and better use of existing available accommodation, but it recognises that there are problems with this sector in terms of some poor conditions and management. The government wishes to retain the good landlords and encourage reputable investors to increase private rented sector supply, whilst making the worst landlords perform better or withdraw from the market – by overhauling local authority enforcement powers (see Chapter 4) – although it is difficult to see how the proposals might achieve a marked improvement in the worst landlords and properties when there is no alternate choice for low-income households unable to afford an alternate choice in the housing market.

Perhaps more likely to have some effect is the proposal to attach conditions to housing benefit, so that the landlord would have to demonstrate decent standards of accommodation and management, although this would require a new form of regulation and, of course, an alternate supply of affordable accommodation for displaced tenants. The government is eager to encourage private rented sector improvements rather than to withdrawal from the market and thus penalising housing benefit tenants. Similar provisions might apply to those claiming benefit, and the government is considering a reduction of benefit to anti-social tenants (DETR 2000). The government's proposals are still at consultation stage, so it is still too early to assess what might happen in detail.

The ways in which the private rented sector can be encouraged to meet housing needs are discussed in Section 5.4.

Summary

- The private rented sector has been in general decline since the turn of the century due to a combination of factors.
- Government policy has continued to favour this sector as providing accommodation for the homeless and those otherwise in housing need.
- The EHO role tends to operate at the bottom end of the private rented sector. Here conditions are poorest and landlords tend to be less scrupulous.
- There are ironies to the new policies – the sector is expensive to live in, is generally in poor repair, particularly to low-income disadvantaged groups, and the housing benefit bill has subsequently grown substantially.
- There is, however, much potential in this sector, particularly in making better use of existing stock and bringing vacant properties back into use for private renting.

2.4 Inequality and poor housing conditions

Outline

Goss and Blackaby (1998) argued that housing policy has always been a dimension of inequality for the following reasons:

- Black, ethnic minorities and woman headed households are over-represented in poor housing.
- Some have more than one home; others have none.
- Some can move freely; others are trapped in negative equity or have an inability to transfer to alternate social housing.
- Rent levels across sector can cause a 'poverty trap' and prevent movement from welfare to work.
- Some poorly designed social housing estates, aggravated by letting policies, have skewed communities and encouraged social exclusion.

There are many complex and interrelated issues making up inequality in housing, particularly evident in recent years. As Balchin (1995: 13) argues, 'Housing policy in the 1980s–mid-1990s, perhaps more than any other issue, was the means by which an increasingly divided society was being created.'

This section looks at the English House Condition Survey (EHCS), issues in housing and disadvantage, social housing provision and UK investment in housing conditions to draw together and explore some of the key issues involved in housing inequality.

The English House Condition Survey 1996

The EHCS is the national survey that has been carried out every 5 years since 1967. It draws together information from across housing tenures and is used to inform housing policy. Reference to the EHCS is also made in Section 3.8.

The most recent EHCS was carried out in 1996 (DETR 1998). It reported that privately rented dwellings are most likely to suffer disrepair, and that the median cost or repair was in the region of £6.50 per square metre. The private rented sector is in a noticeably worse state of repair across all ages. Highest levels of disrepair are found in older urban areas, and up to twice as high in deprived authorities. The estimated expenditure to remedy urgent repairs is approximately £1,280 per dwelling, and this increases by 50 per cent for other repairs, and the figure is rising.

Of all housing stock, 7.5 per cent is unfit. This is the same level as in 1991, and as some have been made fit, while others have become unfit in the intervening period. The highest rates of unfitness are found in pre-1919 stock and converted flats. As a proportion, unfitness is highest in the private rented sector, where 10.3 per cent is unfit as compared with 6.3 per cent of owner-occupied dwellings, 7.3 per cent of local authority and 5.2 per cent of registered social landlord dwellings. The proportion of households living in unfit dwellings has remained the same for owner-occupiers, but has fallen for private sector tenants. This is largely due to the size of the owner-occupied sector. The average cost of making a dwelling fit is £5,230 (DETR 1998).

Of all households, 14.2 per cent live in poor housing, including unfit, substantial disrepair or requiring essential modernisation. Households most likely to live in poor housing are ethnic minorities, especially Pakistani and Bangladeshi as well as black households; the young; the unemployed; older people; households employed part time; lone parents. There has been no overall change in the level of disrepair and unfitness, but conditions for lone parents and ethnic minorities have worsened relative to other categories. Private sector tenants are most likely to live in poor housing, and registered social landlord tenants least likely. Poor housing is erratically distributed nationally, but most poor housing is in old London boroughs, older resorts, and university towns and large urban districts (DETR 1998).

Nearly 1.3 million households are housed in poor living accommodation. This group includes ethnic minorities (19 per cent), Bangladeshi and Pakistani households (30 per cent) and unemployed households (19 per cent). Tenants are more likely to inhabit poor conditions, with local authority tenants particularly being socially or economically disadvantaged. The survey found that some groups of households in poor housing were more dissatisfied than others, particularly young households, unemployed households, lone parents, private tenants, ethnic minorities and households with infants (DETR 1996).

Housing and disadvantage

Since housing is one of the main direct determinants of health inequality (e.g. Townsend *et al.* 1992), it follows that housing policy can be seen as a potential vehicle to alleviate social disadvantage. Finding a suitable definition of social disadvantage is necessary in an attempt to help explain why poor housing and health exist. Disadvantage goes beyond the bounds of poverty, deprivation and inequality, which tend to deal with single aspects and not the combination of people's access to, and remaining in, housing. Clapham *et al.* (1990) usefully identified two broad definitions of disadvantage: the market and social democratic models. Advocates of the market model argue that disadvantage has an absolute scale. They see a minimum set of standards required for subsistence, health and welfare and suggest that continuing deprivation is a consequence of an individual's failure or entrapment in a culture of poverty, and it is up to individuals to resolve it. Their definition looks only at the economic and fails to address social needs. The social democratic model explains how social disadvantage arises and is maintained as part of the organisation of economic and political processes. They then see disadvantage as relative rather than absolute.

Clapham *et al.* (1990) showed how social disadvantage, poverty and inequality are expressed in the housing system. They illustrated how housing is apportioned by a hierarchy of economic and social power that structures society more widely as a dynamic but interlocking set of markets and institutions. Housing is just part of the markets and institutions whereby disadvantage is structured. It is a variety of social, economic, political and demographic attributes that have a bearing on the housing available and its local environment. Thus, housing may facilitate or deny access to wider community and health services. Housing in poor areas is also more likely to have poor local services, preventing disadvantaged individuals from acting fully as citizens. Children born into poor-housing environments are less likely to escape either that environment or the behavioural patterns and lifestyle sometimes associated with it, such as frequent moving (Richardson and Corbishley 1999). This means that the link between disadvantage and poor housing becomes harder to break.

Who then is disadvantaged in housing? Housing in poor condition is not uniformly distributed but is associated with social, demographic and economic factors. It is generally low-income groups who are unable to have an active choice in the housing market (Rhoden 1998). They are, therefore, often marginalised into, or remain in, low-cost housing; sometimes reliant on income-related benefits to do so. Some examples are as follows:

- Long-term older tenants may occupy poorly maintained private rented housing lacking internal amenities, but they tend to have security of tenure.
- Some low-income ethnic minorities have been unable to access owner occupation or social housing due to institutional racism.
- Single people may have less potential to raise a mortgage or secure a market rent without being dependent on benefits and, likewise, other low-income households tend to be marginalised into the lower end of the private rented sector.
- The homeless, frequently requiring other social, educational and welfare support, suffer further disadvantage when placed in temporary accommodation where poor housing standards and community facilities compound their situation.

Policies during the 1980s and 1990s favoured the private rented sector as housing provider. The simple issue of too much demand and too little supply has resulted in increased low-quality housing, increasingly HMO accommodation at the bottom end of the private rented sector, frequently in otherwise undesirable or abandoned areas. With benefits being paid to landlords regardless of conditions, there is little fiscal incentive for landlords to invest in their properties. Deregulation of rents and tenancies has trapped many tenants within the sector long-term, so making tenants increasingly marginalised from mainstream society. This is aggravated by wider economic changes including a general fall in income to the poorest 20 per cent of households due to loss of access to free goods, services and subsidies such as school meals, social fund loans rather than grants and proportional increases in water and local taxation (Townsend *et al.* 1992).

Such issues reinforce the argument that housing regeneration and promoting healthier housing is not just about living accommodation or unemployment. It is also about the complex interrelationship of social exclusion, abandonment of inner-city areas and current lifestyles that make up the bottom end of the private rented sector. The combination effect of these interrelated issues cannot be stressed strongly enough. Housing and health is a two-way street. Access to poor housing generally results from disadvantage and disadvantage frequently results in access to poor housing and therefore poor health.

Social housing provision – meeting need or tenure of last recourse?

Until the 1970s, local authority housing was seen as a 'general needs' tenure, but public expenditure in housing cuts from the late 1970s and political emphasis on owner-occupation began to change this. By 1979, monetarist policies and rolling back the state to extend free market economy and reduce inflation became the favoured political direction of the New Right. One result of this has been a deterioration in housing conditions for low-income groups, especially in the private rented sector and some unsatisfactory local authority estates, with concentrations of some of the poorest tenants and ethnic minorities. Such increasingly strict classification between tenures has caused tension and divisiveness in society (National Federation of Housing Associations 1985).

During this Conservative government, no other specific public service expenditure was so heavily cut back as housing expenditure. Cutbacks in housing expenditure led to decreased local authority building and increasing homelessness, and as central government subsidies reduced, local authority rents had to rise above inflation. Hamnett (1988) offers two explanations for these cuts. The first is as a result of the 1976 sterling crisis and International Monetary Fund pressure, and cuts had to be made to curb inflation and maintain the value of the Pound. The second possibility is that the cuts were ideologically driven by the government to cut local authority expenditure in favour of owner-occupation. This appears the most likely reason since, as Balchin (1995) notes, the net result was a rise in public spending on housing benefit from £280 million in 1979/80 to £3,540 million in 1987/8, and Mortgage Interest Relief rose from £1,639 to £4,850 million in the same period.

UK investment in housing conditions

The UK had the lowest average rate of housing investment in the (then) European Community between 1970 and 1989 (Balchin 1995). This is in part because the UK was already largely urbanised with established housing stock, but there remained a clear need to invest in renewal in ageing local authority and private sector stock. The English House Condition Survey (DETR 1998) simultaneously reported substantial defects across housing tenures, but did not fully address costs of rehabilitation or renewal. It also pointed to private sector housing defects, especially in the private rented sector, the favoured sector for meeting housing need, requiring substantial investment.

Treasury pressure is the main factor in determining housing renewal and the current costs are massive, and can only increase if nothing is done. The nation's housing stock is worth some £3,000 billion. The (then) Department of the Environment in 1985 calculated a cost of £18,000 million to repair or replace defective council housing, but other figures suggest up to £50,000 million is required (Association of Metropolitan Authorities 1985). The private sector required £25,000 million in 1985, but government support was £608 million. Goss and Blackaby (1998) suggest that the figure is now some £50 billion in privately owned stock. However, the alternative to housing investment is neither economically or socially attractive. Without adequate investment, there will be increased housing scarcity, homelessness, rising house prices, social division and decaying areas.

Most poor private sector housing is located in Greater London, which also has the largest numbers of empty houses. At current rates of renewal, London's housing will have to stand for centuries. Like other inner-city areas, many low-income tenants live in poor conditions. There is regional disparity in grant delivery. The English House Condition Survey (DETR 1998) continues to report that owner-occupied housing contains worst housing in absolute terms, but proportionally private rented housing contains most unfits, a disproportionate number of low-income households in worst housing, especially single-adult households, black-owner occupiers and older single people, ethnic minorities in private rented housing. It would take 170 years, without further deterioration, to rehabilitate England's housing stock at current rates (Luba 1991).

The current housing situation is very disparate and its future very uncertain. There are many questions that need answers. Can the social housing sector survive? Is private renting sustainable? What about conditions in the owner-occupied sector? What is the future partnership role for housing associations in providing social housing, private renting (increased funding potential) and the 'gap' in between? How do policies addressing social exclusion fit in with all of this? How will 'single pot' funding benefit the private housing sector? How will housing and disadvantage be addressed?

Increasingly diverse partnerships are becoming key to developing a blend of skills, experience and resources, and alternate possibilities to pool resources of the statutory, voluntary and private sector working together for common aims. But there remains uncertainty in the housing movement and the lack of an overall housing strategy from the government (CIH 1996). There is increased demand for quality housing and services, best value, cost-cutting as well as changing client groups that need to be reflected by the housing provider. There is also a need for specialist skills to cope with, for example, anti-social behaviour and special needs for older people. Partnerships need support from local authorities, health authorities, the police, green groups, specialist agencies, employers and trainers. There needs to be a re-creation of sustainable housing and communities through a balance of economic activities and tenure mix.

Summary

- The EHCS illustrates how some groups are over-represented in poor housing, and that, as a percentage, the private rented sector comprises the worst conditions.
- Disadvantage can be a difficult trap to escape, leading to continued habitation in poor housing conditions.
- There is a need for massive investment across all tenures, with associated policy initiatives, if inequalities in housing are really to be addressed.

2.5 Homelessness

Outline

Put simply, homelessness is about inadequate suitable and affordable housing supply to meet demand. Although it was initially recognised by the Victorians, they did little but stigmatise the homeless and it was not until as recently as 1977 that the first legislation was introduced giving local authorities a duty toward the homeless. Homelessness is largely determined by social, historical, political, economic and investment processes in local and national housing. Constraints in housing provision can arise from a simple shortage of accommodation, to 'interpretation' of legislation and government regulation, and also from the culture of a housing organisation and its organisational arrangements. Policy-making and implementation, subject to suitable housing availability, have become increasingly important following the general shift from the public to private sector housing in meeting need.

Homelessness – a general overview

Homelessness on the increase

There has been an unprecedented increase in homelessness in recent years, displaying regional disparities. The situation is particularly acute in Greater London. Homelessness figures peaked in the early 1990s, but they reduced as a range of initiatives followed. The statistics, however, do not represent the true figure of people without a home at all, or those currently occupying unsatisfactory living conditions. Statistics are confined to those 'accepted' as homeless under law, and many others, such as rough sleepers and some single people, are excluded. Many others still are living in difficult or unacceptable conditions or locations.

There are many reasons for the rise in homelessness and debate around the subject is frequently ideologically driven and value laden. Before the Housing Act 1996, the then Conservative government suggested that there were two main reasons for the unprecedented rise. One was due to changes in behaviour and lifestyle, such as young people leaving home and an increase in illegitimate births. The second was the problem of local authority housing conditions and immigration policies. But perhaps the real reasons were more fundamental in terms of suitable housing supply and demand. Changes in access to benefits (particularly housing benefit and income support), the replacement of urgent needs payment with crisis loans, supplementary benefit being replaced by income support, to be paid in arrears; and a reduction or withdrawal in income support levels to young people by age banding all had some negative impact and role to play in rising homelessness.

Ironically, statistics reveal that there is an untapped supply of housing. The Joseph Rowntree Foundation (1994) reported that English empty dwellings had increased by 42 per cent from 539,000 in 1983 to 764,000 in 1993. There are ten times more vacant properties in the private sector than any other sector. However, there are frequently reasons as to why these are vacant, such as complex reasons of ownership, changes in housing market, lack of funds to renovate, and so on. This is considered further in Section 5.4.

Homelessness – the legal background

The roots of UK homeless legislation can be traced back to the Poor Laws, which distinguished between the worthy and the unworthy poor. The Victorian response to homelessness tended to be one of punishment by way of the Workhouse, rather than assisting those in housing need. It was not until the National Assistance Act 1948 that there was some state provision for homeless households, and authorities were required to provide temporary accommodation for those in urgent need. Homelessness was attributed to personal inadequacy and moral failure and the ideology remained one of anti-State intervention. The State's role was one of judgement and social control, with families often being split, fathers being excluded from hostels and children being taken into care as a direct result of their homelessness, and the idea of housing as welfare need did not enter the equation (Cowan 1999, Conway 2000).

Attitudes toward homeless households began to evolve, and were changed markedly with the screening of the film *Cathy Come Home*, and the subsequent development of the pressure group Shelter. Continued pressure and a recognition of housing shortages led to the government introducing legislation in the form of the Housing (Homeless Persons) Act 1977, which provided an obligation to secure permanent housing for the homeless in priority need (similar to the current definition described below), and to advise others. From its implementation, there were supporters and opponents of the legislation. Some saw it as enabling queue-jumpers to gain permanent accommodation, whilst others suggested that it failed to cater for many of those in real housing need (Conway 2000). The legislation was consolidated into the Housing Act 1985 and was not substantially reviewed until 1992. The subsequent Housing Act 1996 made new provisions for the homeless.

The Housing Act 1996 was in many ways similar to previous homeless legislation in terms of eligibility for housing, and definitions of homelessness, priority need and intentionality. What was new was the way in which the local authority would provide accommodation. The Housing Act 1996 ended a statutory right to access social housing, which would be granted on a waiting list basis. The homelessness was to be seen as a temporary status, so accommodation would be provided as a temporary 2-year solution, normally through the private sector. It was perceived that there should be one waiting list with no priority.

Criteria and definitions

'Accepted as homeless' is a statutory term for those who apply for a local authority for housing as a homeless household and are considered by the local authority to fulfil the statutory criteria set out in the Housing Act 1996. 'Threatened with homelessness' is a statutory definition applied to an applicant who is likely to become homeless within 28 days. Such households are assessed for rehousing in the same way as applicants who are currently homeless.

To be 'homeless' the applicant would have to prove one or more of the following criteria:

- No accommodation with an entitlement to occupy.
- Accommodation with an entitlement to occupy accommodation, but no entry to it.
- Movable structure, but nowhere to locate it.
- Potential or actual threat of violence in the current occupation.
- Unreasonable to continue to occupy the current occupation.
- Accommodation is not available to the whole household and anyone else who normally lives with them.

The local authority has a statutory duty to accept an applicant as homeless if they are:

- A 'priority need', which is defined as including pregnancy, having dependent children, vulnerable, e.g. older people, those with learning or physical disability, homeless or threatened with homelessness due to emergency.
- Unintentionally homeless.

- Eligible for housing assistance.
- If there is no other suitable accommodation in the area.
- If there is a local connection (e.g. was or is living in the borough).

In addition, local authorities are required to give reasonable preference to persons who:

- Occupy insanitary or overcrowded houses.
- Live in unsatisfactory conditions.
- Live in temporary or insecure accommodation.
- Have dependent, or are expecting, children.
- Have medical or welfare needs.
- Have social or economic circumstances leading to a difficulty in securing settled accommodation.

Cowan (1999) argues that there are major differences in the way in which many of these definitions are interpreted, deliberately or otherwise, which results in many households in real need being excluded from entitlement. Many households in housing needs are not accepted as homeless by local authorities because they do not meet the eligibility criteria and there are no statistics available on this group, which includes people in unsatisfactory accommodation, travellers, women with violent partners and people sleeping rough. Many people move from one insecure situation to another, particularly young people who have been in care, those with mental health problems, refugees and asylum seekers, ex-members of the armed forces, and so on, who may not be eligible for social housing (Conway 2000). Authorities also have certain obligations for those in need under the Children Act 1989 and the National Health Service and Community Care Act 1990. An example of how a local authority might determine an application of someone presenting as homeless is shown in Box 2.1.

Homelessness and temporary accommodation

A definition of temporary accommodation is provided in Section 2.2 (homelessness and temporary accommodation), and should be read in conjunction with this section. Temporary accommodation was seen by the (then) government as one way forward in addressing rising homelessness levels under the Housing Act 1996. It was also part of its ideology of introducing a market into (social) housing provision, packaged as an element of choice, rather than as insufficient local authority supply (e.g. Rhoden 1998). Others were not so optimistic, arguing that that temporary accommodation would exacerbate the situation; increase financial and other costs; lead to a decrease in social house building; and have an

Box 2.1 Theory into practice – intentionally homeless or not?

Local authorities need to investigate fully all applicants presenting as homeless before making a decision on whether they are homeless, eligible for assistance and are satisfied that the homlessness is non-intentional.

A family applied as homeless, and the local authority found a priority need because of having dependent children and thus provided temporary accommodation for a reasonable period whilst obtaining relevant information to consider the application further. Investigating the situation further, the local authority found that family was in rent arrears. This raised the possible issue of intentionality – a person is deemed intentionally homeless if they do or fail to do anything that leads to their right to occupy accommodation, a common example being someone who deliberately chooses not to pay rent.

The family's situation was investigated and it was found that it had not paid its rent because the members were not happy with their allocated tenancy, and they thought that the arrears were not their problem because they received housing benefit. The family had other outstanding debts, not just rent arrears, arising from a low fluctuating income and was sometimes unable to place rent payments above more pressing debts. The family had been evicted because of rent arrears and was deemed intentionally homeless by the local authority. In such cases, the local authority normally serve a section 64 notice, which is a declaration not to offer rehousing because the homelessness has occurred through intentional non-payment of the rent.

The result was that the local authority would not rehouse the family, which had to find its own accommodation. This compounded the debt problems the family faced, but there was little the local authority could do but refer the family to someone who could assist in giving money and debt advice.

If the same family had not been in rent arrears, but had lost its privately rented home because, for example, it was repossessed when the landlord had not paid the mortgage, the family could have been accepted as non-intentionally homeless and in priority need, and so be rehoused by the local authority.

overall result in a skewing of demand rather than addressing and providing for real housing need.

Temporary accommodation is made up of hostels, refuges, bed and breakfast hotels, short-life housing and leased private sector dwellings. Local authorities place homeless households in temporary accommodation for several reasons, which Conway (2000) identifies as including:

- Whilst waiting for full investigations into the household's circumstances before a formal decision can be made.

- Owing to a shortage of suitable permanent accommodation.
- For use as a deterrent to those who may be less genuine or urgent.

Temporary accommodation is frequently unsatisfactory on many levels, both in physical condition as well as stress to the occupiers. A stay in temporary accommodation may last for an extended period and may not be in a convenient location. Whilst such accommodation cannot be the answer to the nation's housing crisis, local authorities recognise that, in the absence of other available accommodation, they need to ensure that standards are as reasonable as possible. It is important that local authorities have organisational procedures in place to ensure that the temporary accommodation they are using meets required legal standards, and this is often achieved with housing officers and EHOs working closely together.

The use of temporary accommodation rose through the 1980s and peaked in 1992, but it reduced following a range of initiatives, as illustrated in Table 2.2, not least an increase in supply of housing association properties. However, as Fox (2000) reports, the number of households in temporary accommodation reached an all-time high in Summer 2000. At the end of June 2000, 66,030 families were waiting to be permanently housed by councils, a rise of 4 per cent on the previous quarter and 11 per cent on the same quarter the previous year. There was also a rise of 2 per cent in the numbers in bed and breakfast accommodation, a total of 8,380 and 9,530 in hostel accommodation. Shelter argued that these rising figures were due to low-income groups, particularly in London, not being

Table 2.2 Initiatives to address rising homelessness

Date	*Initiative*	*Purpose*
1989	Homeless Initiative	Mainly to renovate local authority and housing association accommodation and to lease private sector housing, with grants to the voluntary sector
1990	Rough Sleepers Initiative	Provision of hostel spaces and permanent move-on accommodation
1991	Empty Homes Agency	Establishment of the Empty Homes Agency to bring the UK's vacants back into habitable use
Later schemes	Various national and local schemes, notably LOTS, housing in repossessed accommodation, private sector leasing, etc.	To encourage greater flexibility in housing provision, and, in particular, to make better use of the private sector

able to afford high rents or access mortgages and it suggested that 100,000 new affordable homes should be built between 2000 and 2011. Shelter reiterated that temporary accommodation is totally unsuitable for family life.

There are many ironies in levels of homelessness and policy response that can only really be addressed by reference to government ideology. Whilst levels of homelessness were rising at unprecedented levels in the 1980s, and peaking in the early 1990s, government response was to reduce the level of social housing units provided by local authorities, both due to council house sales and reduction in new builds. Homeless households comprise the majority of local authority lettings and increasingly of housing association nominations. Conway (2000: 80) argues that 'If local authorities had still been building and letting on the same scale as in the 1960s and 1970s, these same families would have a chance of rehousing from the normal waiting list before their situation became urgent.' Balchin *et al.* (1998: 54) support this view, adding of the policy irony that 'as the number of homeless reached record levels in the early 1990s, half a million building workers had been made redundant in the worst slump in the construction industry since peacetime since before 1914.'

The costs of temporary accommodation are social as well as economic. The use of temporary accommodation generated substantial diseconomies in public expenditure, with the bill shifting from housing to social security expenditure (Balchin 1995). The National Audit Office (1989) estimated the cost of a family in bed and breakfast accommodation to be £15,500 annually, whereas new local authority housing would cost £8,200. Accommodation is generally poor standard and there is a lack of choice. There are issues of physical and mental ill-health arising from uncertainty and conditions (Rhoden 1998, Stewart 1999, Conway 2000). Such 'privatisation' of housing is bad value for money and for people's lives.

Such arguments led to increased use of private sector renting and leasing, which is discussed further in Section 5.4. Whilst this is preferable to some other forms of temporary accommodation, it can be difficult to afford because rents are not always covered by housing benefit, and the length of tenancy remains insecure to the occupier, causing stress. There are many different forms of arrangement with the private sector to provide housing, including leasing and various forms of management by housing associations and others (Conway 2000). However, many local authorities have no option but to look to the private sector to meet need, because they simply have inadequate low-cost housing stock of their own and resources remain inadequate to meet growing demand.

There is also the issue of those a local authority cannot, or will not, accept as statutorily homeless. Many fall into this category and may be occupying unsatisfactory living accommodation, or may have no home at all. Voluntary organisations such as Shelter's Crisis seek to support such individuals, with much success in securing sustainable housing and meeting other social needs. With mounting pressure to address street homelessness, there has been a range of government initiatives in recent years (Table 2.2), including the recent appointment of the 'Street Tsar' to identify initiatives to address street homelessness.

Summary

- Arguments remain that housing the homeless is value-laden and ideologically driven, dividing the homeless into categories of deserving and undeserving.
- Homelessness is fundamentally about supply and demand. When supply is short, legislation has responded by altering the definition of 'need'.
- There has been a range of recent initiatives to address homelessness, many of which are increasingly concerned with enabling housing provision through the private sector, but with a lack of affordable social housing, numbers of homeless households continue to rise.

2.6 Asylum seekers and housing

Outline

The Asylum and Immigration Act 1999 removed asylum seekers from mainstream welfare support, which is replaced with a totally separate system of cash-less support and has made new provision for housing. The intention of this new policy seeks to discourage 'economic migrants', defined as those seen to relocate for financial benefit, rather than those who have genuine grounds for seeking asylum. There can be little doubt that the media interest has had considerable, often negative, impact on attitudes toward asylum seekers in the UK. This section first defines an asylum seeker before looking at emerging policy issues surrounding their housing and support.

Asylum seekers and housing opportunity

Defining immigrant, asylum seeker and refugee

An immigrant can be defined as someone who moves from his or her homeland to a new country and the term is often associated with someone who migrates in search of work, usually from a rural to an urban location. The move is voluntary, unlike that of a refugee, where the move is forced. A refugee normally has the rights of a permanent resident including the right to work and seeking benefits under the 1951 United Nations Convention on Human Rights. The terms 'immigrant', 'guest worker' and more recently 'asylum seeker' are indicative of the status of the individual in the host country as distinct from the indigenous population. The terms imply temporary residence, and having only limited rights (Bullock *et al.* 1988). An asylum seeker is defined as someone who comes to the UK, often fleeing persecution, torture or war, and who applies for refugee status. A person remains an asylum seeker, with limited rights, whilst the Home Office considers their application, and any appeal.

New powers – Asylum and Immigration Act 1999

The Asylum and Immigration Act 1999 introduced new powers and controls to asylum seekers awaiting decisions on their Home Office applications. The Act removes asylum seekers from mainstream support and benefits previously provided under the National Assistance Act 1948. This is replaced with a discretionary system of benefits in kind (CIH 2000). The overall intention is to discourage 'economic migrants'. Numbers seeking asylum have risen markedly in recent years, due largely to political instability in some regions, and some of the media have generated negative hype and a discourse controlling debate about asylum and immigration, frequently without empirical evidence (Cowan 1999, Gilliver 2000).

Provisional data from the Home Office (2000) shows that around 6,000 people applied each month in Summer 2000 for asylum in the UK, which is a slight decrease from the same period in the previous year. The asylum seekers are a mixture of nationalities, but mainly from Iran and Iraq. Numbers of applications at ports have remained fairly level, but have increased in-country and by postal application. The proportion of cases recognised as refugees and granted asylum has remained relatively consistent at 9 per cent, with those granted exceptional leave to remain being around 12 per cent. There has, however, been an increase in the number of appeals to the Home Office. The Local Government Association estimates that there may be 2,600 asylum seekers cases per month requiring dispersal in the UK (LGA 1999).

Gilliver (2000) notes that:

- The UK is ninth in Europe in terms of asylum applications per 1,000 inhabitants, receiving less than one-third of the number received in Germany.
- Over half of asylum decisions in 1999 led to permission being granted.
- Asylum seekers actually receive 70 per cent of income support, paid largely in vouchers.

Accommodating asylum seekers

The National Asylum Support Service (NASS) is the new executive agency of the Home Office charged with administering the new system. NASS's remit for assessing and providing support outside of mainstream welfare services, with the exception of those with transitional or continuing rights who can access some welfare benefits and local authority services. Its main objective is to oversee a voucher-based system, with a small amount of cash weekly, and dispersal of asylum seekers to reduce pressure in London and the South East. At present, the NASS is only responsible for port applications in England and Wales, but once the Act is fully implemented – expected to be in April 2001 – it will be responsible for all UK asylum seekers. Until then, under the interim scheme, local authorities will retain the duty to assist destitute asylum seekers. Local authorities can discharge this duty through another local authority and be reimbursed with fixed rate government funding (CIH 2000, Resource Information Service 2000).

The interim and NASS schemes seek to reduce pressure on port authorities and London by 'dispersal' programmes. Asylum seekers needing accommodation will be dispersed in 'cluster areas' to create new ethnic mini-communities (CIH 2000), with the intention that this will help provide mutual support and to deter migration to existing overburdened locations. Under interim arrangements, this is voluntary under Local Government Association guidance (LGA 1999).

NASS will be responsible for securing accommodation and support for asylum seekers, on a contracting basis at flat-rate financing. NASS will contract with suppliers across all tenures, to supply accommodation and services at a fixed sum, which covers all expenses. Landlords would be expected to cover all their charges, services and expenses in the contract price (LGA 1999) as follows:

- Rent.
- Full furnishing.
- Designated play area.
- Communal rooms.
- Areas for language classes.
- Créche facilities.
- All utility costs.
- Additional security costs.
- Staff costs, including a unit manager and staff.
- Scheme translation costs.
- Other administration or management costs.

The support for destitute asylum seekers is based on the assumption that those provided with accommodation through NASS, with associated funding for utility supplies, will only require 70 per cent income support level rate for essential living expenses. Support is in the form of vouchers for exchange of goods and some of which can be cashed in, but change from the vouchers is not permitted. Asylum seekers choosing to live with friends or relations would not be eligible for payment.

Asylum seekers needing accommodation would be dispersed to cluster areas, based on the Home Office criteria, cited in CIH (2000), of:

- Availability of suitable accommodation.
- Existence of multi-ethnic population or support services for assisting asylum seekers.
- The potential to link in with existing communities.
- For these reasons, NASS takes account of:
  - the applicant's cultural background and first language;
  - the applicant's accommodation needs (family, disability);
  - characteristics of the locality; and
  - accessibility of support services within reach by public transport.

NASS has no scope for the applicant's personal preference and through the scheme is exempt from tenancy legislation. Once an asylum seeker is awarded full convention status or exceptional leave to remain, they would be required to move on within 14 days.

When responsibilities are transferred to NASS, its is hoped that 40 per cent of accommodation will be contracted out to nine local authority consortia, but only one, Glasgow, has signed up so far (Albeson 2000, Hawkey 2000). The remaining 60 per cent will be contracted to the private sector. There is currently some supply through the private sector, including private residential companies as well as some ex-local authority stock (Hawkey 2000). However, the supply is only one issue. The fact that there is over supply in some areas may be because no one else will live there or that conditions are unsatisfactory (Albeson 2000). The second point is the real need for good, sensitive management and provision of additional, unique services required by asylum seekers – as well as a decision on who is expected to provide these services.

Social-type housing management, and indeed housing asylum seekers or others with specific needs, has not been a traditional function of the private housing sector, and there is a very real risk that this could be overlooked in a contracting climate where cost is key. Despite the need for appropriate private sector housing management for asylum seekers, at the time of writing there has been no Home Office guidance. Some in the social housing sector have expressed concern about whether the private sector can or will really fill the social role required in housing asylum seekers, other than to meet a minimum contract requirement (Hawkey 2000). NASS will consult with the relevant local authority when the private sector is to provide in its area, so it can maintain a strategic role.

Refugees

An asylum seeker receiving refugee status and associated rights, including the right to work, receive benefits and apply as homeless where appropriate. Whilst they can apply as homeless and claim benefits, they may need support to do this or might risk losing that entitlement due to language difficulties or remaining traumatised at their situation. The normal homeless legislation would apply, and their application would depend on assessment of priority need or otherwise. Many stay in temporary accommodation for up to 2 years (Albeson 2000) before being rehoused.

Asylum seekers and housing conditions – the environmental health role

Asylum seekers tend to be concentrated in low-quality, high-risk bed and breakfast-type establishments and local authority environmental health departments have played a pivotal role in enforcing and advising on housing standards for some time. Although the system of asylum support is set to change, it seems likely that asylum seekers will still be forced into housing at the bottom end of the private rented sector. Asylum seekers are already disadvantaged in society and excluded from participating in the housing market, aggravated by a continued withdrawal of benefits, and now being removed from that system altogether. Whilst some of the new changes are beneficial in requiring standards of the proposed housing providers, it is difficult to assess where the accommodation might be and if it can really herald a move away from current provision.

Asylum seekers arriving in the UK have acute housing needs that require immediate assistance. There are many refugee support organisations that place significant numbers of homeless asylum seekers in the private rented sector and provide support services, such as interpreters. What happens to asylum seekers, their numbers and location is not always known to the local authority. Unfortunately, the accommodation they are placed in is not always satisfactory and sometimes it is dangerous, or it is unknown to the environmental health department responsible for ensuring adequate conditions. There are specific health risks in this sector, including tuberculosis. The EHO role is, therefore, pivotal – already providing an integrated response to housing and communicable disease control, as well as an input into anti-poverty strategies and the new public health (see Section 2.2).

There are several issues that make asylum seekers' housing more complex. Already disadvantaged and often alone, their needs are wider than just housing and they require specialist advice and support services – including Home Office restrictions on moving house that may result in benefit being discontinued. There is a shortage of accommodation and needs to be a balance between placing asylum seekers in satisfactory accommodation, their being able to stay in that accommodation and having adequate support agencies to meet their unique needs. Sometimes environmental health activity results in eviction, meaning that the asylum seeker may lose their right to benefit. Clearly this is not desireable when low-cost accommodation is already in short supply and partnerships (see the definition in Section 5.6) are addressing this situation (Page 1996).

Housing for asylum seekers presents a wide range of problems, so it requires a wide range of solutions. There are many statutory and non-statutory organisations with a role to play and local authorities, particularly in London, have been working in partnership for some time to address the poor living conditions many asylum seekers face. They have identified many issues common to asylum seekers and housing and see their role operating best in partnership across statutory and non-statutory organisations, that asylum seekers regularly approach for help. This will bring together resources and help develop a more streamlined and effective service both to provide initial resources (such as blankets, etc.) as well as ensuring safe and secure living accommodation and meeting social needs.

A London-based conference was held in 1996 to bring together key organisations and consider a way forward (Page 1996). Recommendations of this conference can be largely summarised into the following points:

- Development of partnership approaches that are workable and provide support and guidance on rights for asylum seekers on arrival – to include local authority environmental health and housing departments, refugee advice groups, health organisations, etc.
- Working groups of field staff to meet regularly to ensure a flow of information and coordinate action to develop best practice.
- Planned, proactive inspections of HMOs based on good practice guidelines to avoid loss of lettings.
- Joint meetings to establish areas of overlapping responsibility and training needs.
- Establishing a database of HMO landlords across authorities so that asylum seekers can be directed to the most suitable accommodation and liaison with neighbouring authorities (similar to BABIE).

- Look to longer-term housing supply, e.g. leasing, housing association nominations, HMO accommodation that meets minimum standards.

Members of the CIEH London Centre's housing study group are actively working to raise the profile and have recently reiterated that there needs to be close liaison between services, that minimum standards should be agreed on a 'home authority' principle and that the Local Government Association should develop good practice guidelines (Hatchett 2000).

NASS is set to bring major changes to the way in which asylum seekers are housed and supported, and there is scope for building on existing local authority and voluntary partnerships to promote the living conditions and resource needs of this specifically disadvantaged group.

Implications

The implications of these changes are still unknown and it is too early to assess what might happen longer term to asylum seekers' housing opportunities. Those who fail to secure social housing would be likely to find themselves in the lower cost end of the private rented sector and there is already evidence that asylum seekers are living in poor bed and breakfast accommodation. There can be little doubt that poor housing combined with an asylum seekers' trauma of their immediate past, language and cultural barriers, being in an unfamiliar environment, anxiety and so on would aggravate ill health (CIH 2000). There is a 2–3-year waiting time for asylum application to interview, which the Local Government Association has estimated at a cost of £3,000 million this year (Hatchett 2000).

Summary

- The Asylum and Immigration Act 1999 replaces asylum seekers' previous access to benefits with a voucher-based system, administered through the NASS.
- There is a risk that the new legislation will result in labelling of, and discrimination against, asylum seekers.
- The rising numbers of asylum seekers requires good, proactive management.
- Asylum seekers have specific housing and community needs with appropriate support services.
- The private sector is expected to meet 60 per cent of housing provision, but there are concerns about the level of support this sector will provide in a contracting environment.
- Partnerships continue to promote the profile of asylum seekers and promote positive change.

2.7 Fuel poverty, energy efficiency and conservation

Outline

The debate around fuel poverty (as defined below) has been guided perhaps as much by the wider environmental movement as by fuel poverty in itself. By the late 1980s, there was much debate about environmental and social issues connected to energy and emerging green policies. Issues of energy conservation attracted attention, including associated issues of tackling fuel poverty, health and housing inequality, linked to energy consumption and the role housing has to play in environmental change. The legislation emerging from this movement provides the duty for local authorities to address energy conservation, which would inevitably have some impact on fuel poverty. This section should be read in conjunction with Section 4.17, which considers the Home Energy and Energy Conservation Acts.

Toward energy efficiency

The United Nations Commission on Environment and Development (UN 1992) set out ways to work toward sustainable development (see also Section 2.2). This can be defined as development meeting the needs of the present, without compromising the ability of future generations, to meet their own needs (Sustainability 2000). The Rio Summit provided agreement to reduce 'greenhouse gases' to 1990 levels by 2000, including a commitment to reduce domestic energy consumption and emissions. The practical implementation of this policy is based largely on local authorities that are the strategic enablers and coordinators of the Home Energy Conservation Acts and Local Agenda 21, which focus on the built environment for domestic energy standards.

Domestic energy consumption accounts for 30 per cent of total fuel combustion and a similar proportion of energy-related carbon dioxide emissions. Cold homes represent the primary health risk associated directly with housing stock conditions, but some improvements can be made relatively simply and inexpensively (Table 2.3). Two factors generally arise: energy conservation and affordable warmth.

Summary of the English House Condition Survey Energy Report

The first English House Condition Survey Energy Report was carried out in 1991 (DoE 1996) and suggested four key reasons behind inadequate energy conservation standard in the nation's housing: physical standards, energy consumption, thermal conditions and energy action.

Table 2.3 Energy saving in the home

Area of home	*Heat loss (%)*	*Solution*	*Examples*
Walls	35	insulate	cavity wall insulation
Roof	25	insulate	loft insulation
Floor	15	insulate	under-floor insulation
Window/door	10	insulate	double-glazing, draught excluders
Draughts	15	draught proofing	block unused chimney, draught proof ill-fitting windows and doors

Source: based on NEA (undated), Newcastle

Physical standards

Energy rating is usually measured by the Standard Assessment Procedure (SAP). It is based on heating facilities and thermal insulation. Energy efficiency and household standards vary with tenure, dwelling type, region and location. Heat loss through inadequate insulation is very high and most houses fall below current requirements. Pensioners and lone households tend to have less energy-efficient homes than families and young couples.

Energy consumption

Legal standards for heating are set out in the Housing Act 1985 (as amended) fitness standard and are very low. The standard is normally interpreted as requiring a fixed heater in the main living room and provision for heating (a socket) in each other room capable of achieving specified temperatures in given times. Heating depends largely on lifestyle and is often inadequate. Types of fuel used vary widely and 'affordability' is key. Fuel expenditure varies from 5 to 23 per cent net household income; the average being 8 per cent.

Thermal conditions

Causal factors of poor thermal conditions include the heating regime and 'affordability'. Only 25 per cent of homes meet minimum statutory standards for new homes. Damp and mould growth are frequently interrelated, particularly in the private rented sector. The type of heating used tends to correlate with income.

Energy action

Change can be measured in percentage savings in the cost of heating. There has been massive investment in recent years on energy-related work, but a general increase in household appliances. Attitudes to insulation work vary; owner-occupiers tend to do minimum work and tenants even less. Although there would be improved potential, works would be expensive and lead to increased comfort rather than energy savings.

The past 20 years have seen a massive increase in central heating, paralleled by improvements in energy efficiency, so fuel consumption has remained relatively static. There is still a need for greater 'affordable warmth' policies for vulnerable households and opportunities exist through various organisations to promote this. Such organisations include the Home Energy Efficiency Scheme (HEES) and electricity and gas suppliers, and the voluntary sector. Key findings of the EHCS Energy Report (DoE 1996) were:

- Forty per cent of privately rented homes, 20 per cent of socially rented homes and 10 per cent of owner-occupied homes are grossly energy inefficient.
- One in five homes are incapable of being made energy efficient.
- Five per cent of homes overall are energy efficient (with SAP ratings of 60).
- Energy consumption has generally increased.

Fuel poverty

Fuel poverty is the inability to achieve affordable warmth mainly due to poor levels of energy efficiency in the home. An increase in income alone is not sufficient to alleviate fuel poverty, but financial hardship does play a role. Poor insulation and heating systems are the main cause (Table 2.4).

The extent of fuel poverty in the UK is discussed in Boardman (1991) and includes the following issues:

- Thirty-one per cent of UK households are in fuel poverty, which accounts for more than 7 million households, with 'poverty' being defined as eligibility for income-related benefits, many of whom have no capital at all.
- Sixty-eight per cent of fuel poor live in rented accommodation, with no legal rights to carry out works, 27 per cent of whom are older people.
- The fuel poor include pensioners, the unemployed, the sick, and the disabled, who are home all day, requiring 13 hours of affordable warmth.
- Poor households tend to have lower average temperatures than higher income households who have centrally heated households.
- Coldness can lead to dampness.

Table 2.4 Cold houses – reasons and remedial action

Reason	low income	fear of high bills	poor thermal insulation	poor heating system	misuse of heating controls
Action	financial advice to maximise income, including benefits	advice on budgeting, paying for fuel and costs	advice on insulation and possible grant	advice on improvements or new system	advice on using heat controls for energy efficiency

Source: based on NEA (undated), Newcastle

- Many die annually from cold temperatures, this level being more in the UK than in many other colder countries.

There are well-established health risks arising from cold temperatures (NEA 1996). Healthy and active people can generate more of their own heat at colder temperatures than sedentary or sick people. A living room temperature of 18–21°C is 'comfortable', but exposure to lower temperatures, especially if prolonged, can result in:

- at less than 16°C, a marked decrease in the body's ability to stave off respiratory illness;
- at less than 12°C, an increased blood pressure and heart rate, possibly leading to other cardiovascular disorders; and
- at less than 6°C, a risk of failure of the body's sensitivity to temperature change, leading to hypothermia.

The local authority role in dealing with the mixture of energy efficiency and fuel poverty is wide ranging across all housing tenures. It is encompassed in strategies such as Local Agenda 21, anti-poverty strategies, home energy efficiency local energy conservation policies and enforcement of housing legislation. This is considered further in Section 4.17, which considers the Home Energy Conservation Act and the Energy Conservation Act, as well as the Home Energy Efficiency System. Activity should be reported through the annual HIP bid, illustrating action taken and action proposed in respect of domestic energy efficiency.

The local authority role has involved collating existing information on energy efficiency concerning the authority's own stock, on which information is readily available, but information on private sector stock requires further research. The way in which local authorities have set about their task has differed from area to area. Some have undertaken thermal imaging, whilst others have taken more of an administrative role in scrutinising archived records for energy information. The local knowledge of officers is key, such as recalling details on heating and insulation facilities in HMOs and so on. Computer databases need to be able to support required information and should be upgraded accordingly. Some authorities have appointed an individual to the task whilst others have been required to absorb the task into their existing workload. No extra funding has been made available, so local authorities have had to be innovative in making better use of existing funding from discretionary grants and HEES, as well as forging links with the private sector, for example in providing free energy-efficient light bulbs and low-cost, low-energy refrigerators in the manner described in the local house condition surveys section. There is scope to disseminate information and advice through landlords' and tenants' forums.

See also Section 4.15, which discusses the Home Energy Conservation Acts 1995 and 1996.

Summary

- Fuel poverty, energy efficiency and conservation are closely related issues.
- Many local authorities have active programmes in place to tackle fuel poverty across all tenures, but there are few enforcement powers to require improvements in the private sector, so education and advice are paramount.
- The Home Energy Conservation Act and Energy Conservation Acts require local authorities to cut domestic emissions of greenhouse gases through a combination of advice, education and partnership working – but with no additional funding.

Chapter 3

Essential basic knowledge

This chapter is more technically based than Chapter 2. It is a 'how to' rather than a 'why to' and discusses essential basic knowledge before detailed environmental health housing law and practice is presented in Chapter 4.

Chapter 3 introduces essential basic knowledge required as a background to deal with poor housing conditions. It provides a foundation to building terminology and common housing defects before using this as a basis on how to survey a house and draft a specification of works. The chapter considers key legal terminology required in applying housing legislation in respect of condition and tenancy; introduces private sector housing finance and overviews concepts in house condition surveys. It is summarised in a section on basic issues in housing practices and procedures, which includes proper service of notices, prior to Chapter 4, which is concerned with the application of specific environmental health housing law and practice.

The chapter is structured as follows:

3.1 Building terminology
3.2 Common housing defects
3.3 Surveying a house
3.4 Drafting a specification of works
3.5 Key legal definitions for housing enforcement
3.6 Introduction to landlord and tenant law
3.7 Basic private sector housing finance
3.8 House condition surveys
3.9 Housing practice and procedures

3.1 Building terminology

Outline

Knowing key building terminology is by those involved in housing enforcement and related areas, such as housing grants, for several reasons. Fundamentally, it is the basis of identifying building defects in deciding action to take and drafting accurate schedules of work. A sound knowledge is vital in displaying professionalism to builders, architects, surveyors, landlords and so on. This section introduces key building terminology through line diagrams backed up by some discussion of relevant issues.

Key building terminology

Whilst some terms are the same nationally, others can be regional, depending on local building materials, the history of the area, etc. The following is a general guide to common terms, accompanied by diagrams indicating specific parts, but the list is not exhaustive.

Brickwork

There is an infinite amount of brickwork available, changing on a regional basis, and specialist information should be sought in each circumstance.

Pointing is the sand and cement between the brickwork, which can be provided in various styles to disperse rainwater.

Windows

Casement windows can be wooden, metal or uPVC and are hung by hinges, being top hung or side hung casements. The casement is the openable part (Figure 3.1).

Double-hung sliding sash are traditional type windows where the two sash windows move up and down across each other by weights enclosed in the frame (Figure 3.2).

Window-sills should be positioned to overlap the wall, with a 'drip' to prevent water running back inside the premises.

Doors

The door and frame is collectively known as the doorset. Traditional doors tend to be made of top, middle and bottom rails, with vertical stiles (Figure 3.3). More modern doors are flat and called 'flush'. The door shuts against a doorstop. The door is framed by an architrave. The door handle, latch and lock should work effectively. Weatherboards to external doors should have a drip to the underside to prevent water running inside the premises.

Fire doors (and sets) provide at least 30 minutes' fire integrity and are referred to in Section 4.11.

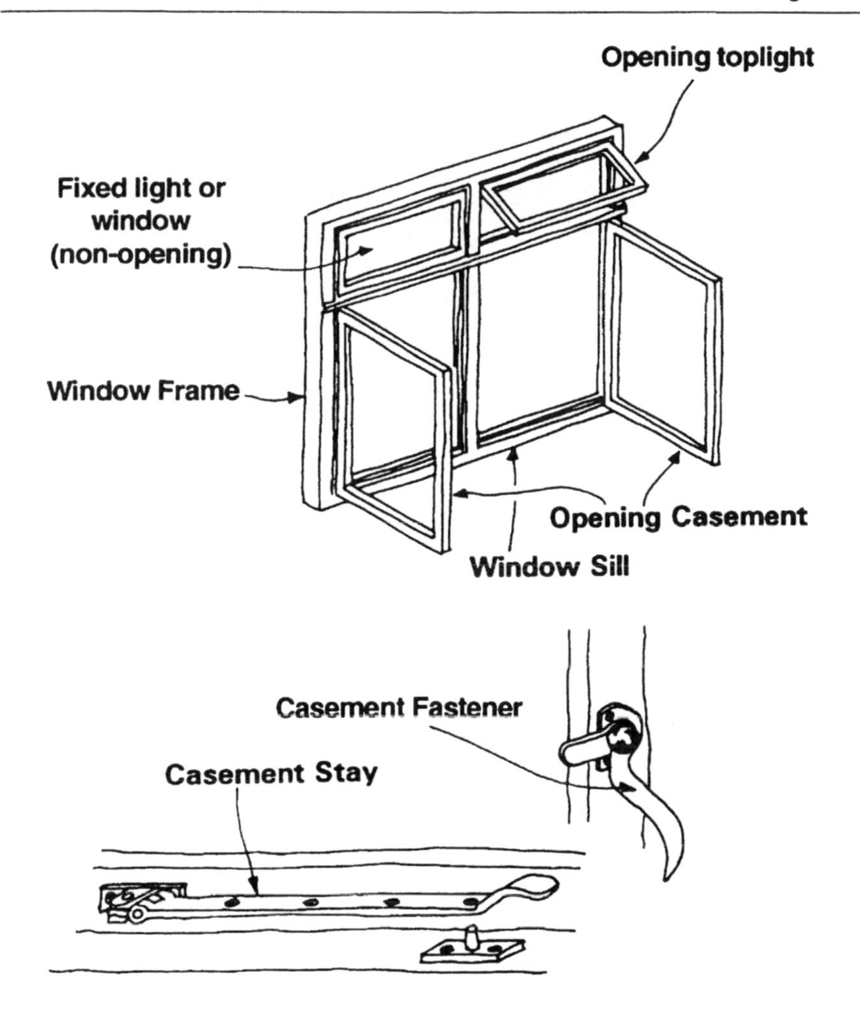

Figure 3.1 Parts of a casement window

Doors and windows are fitted into the reveal of an opening (Figure 3.4), with the jamb at its corner. The wall above the opening is supported by a lintel, which is a horizontal timber or stone providing tensile strength.

Roof

The purlin is the horizontal beam resting upon the walls and supporting the rafters or roof joists. The rafters, which are then felted and battened, support the slates or tiles. Slate roofs displaying nail fatigue are evident

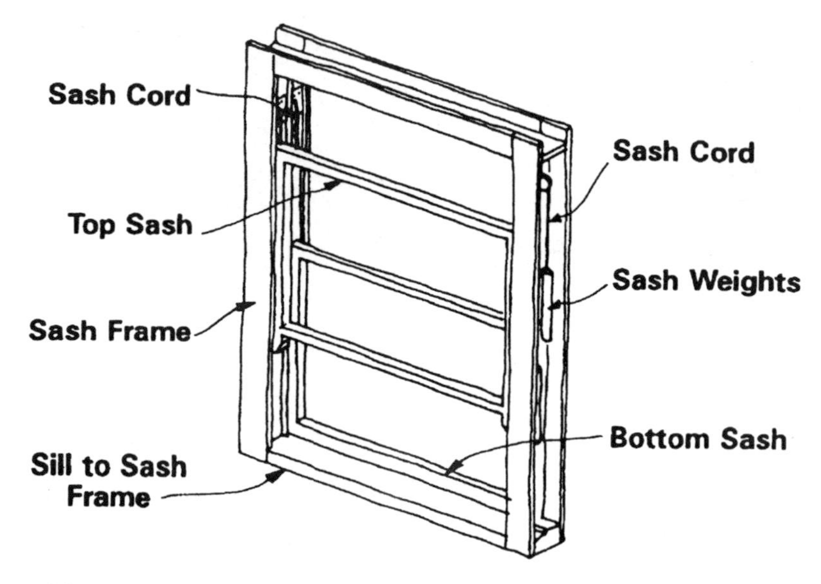

Figure 3.2 Parts of a sash window

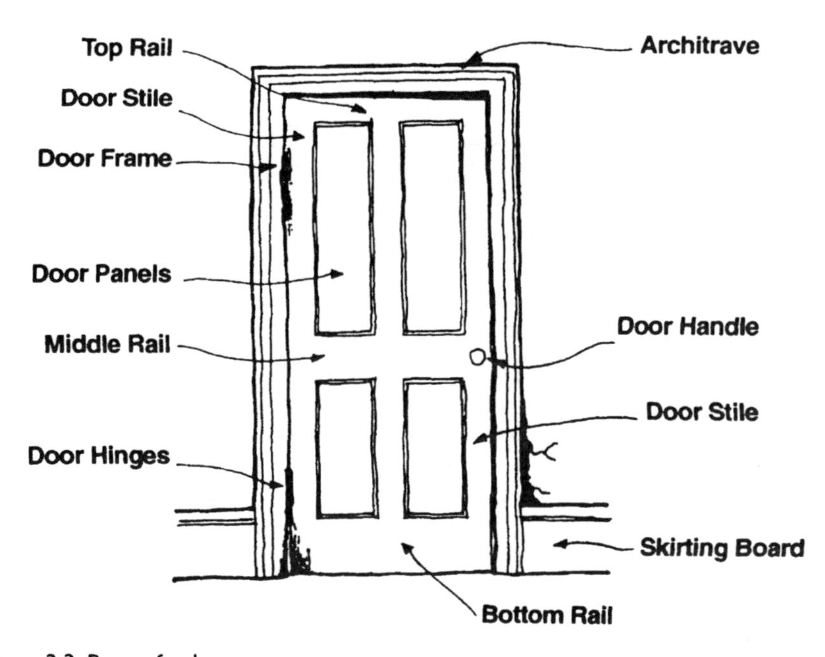

Figure 3.3 Parts of a door

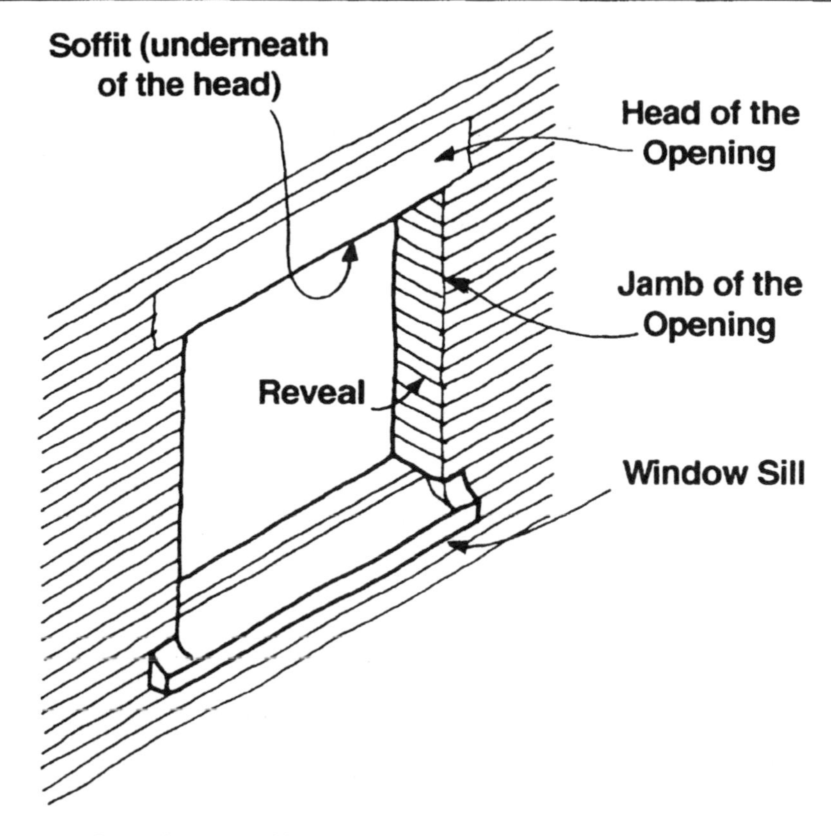

Figure 3.4 Parts of a window/door opening

where metal hooks overhang the battens and support the slates. Parts of a roof are shown in Figure 3.5.

Ridge tiles are shaped tiles that cover the ridge (the top) of the roof. Hip tiles are shaped tiles that cover the hip (diagonal side) of the roof. The verge is the overhang of the pitched roof beyond a gable wall. Fascias are the vertical boards (normally wood or uPVC) attached to rafters or bearers, edging the roof at the eaves and supporting the guttering. The soffit is the horizontal board underlining the rafters.

The chimney pot sits on the chimney to ensure that smoke is dispersed away from the premises. Flaunching is the mortar holding the chimney pots in place on the chimney stack. Flashing is situated between the junction of adjoining structures, such as between chimney and roof or between wall and roof. Stepped flashing is fixed between the brickwork and is held in place by pointing (the cement between the brickwork). The apron extends over horizontal areas of roofing between the

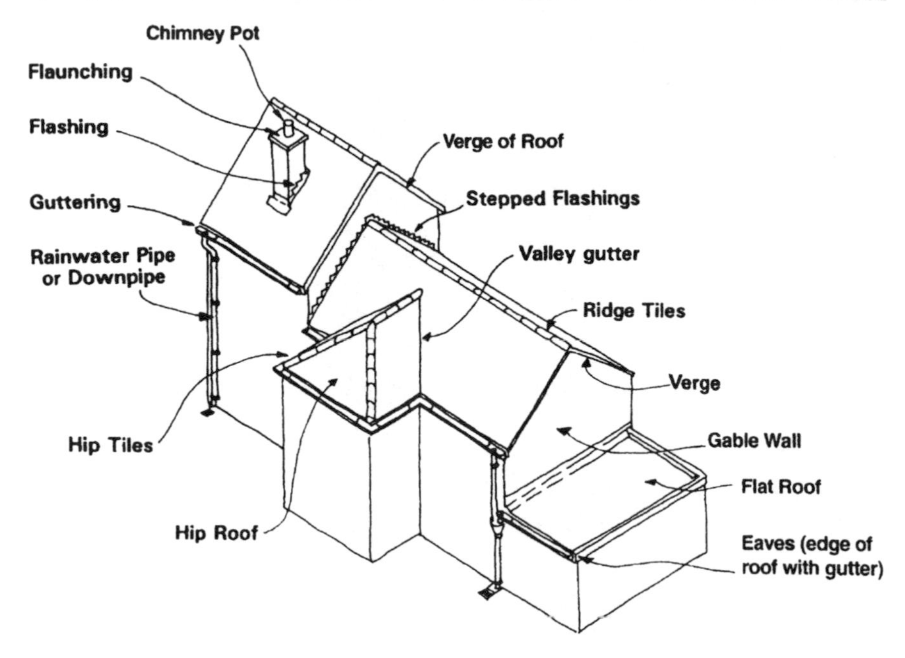

Figure 3.5 Parts of a roof – pitched and flat

chimney and roof slates or tiles. Soakers serve as roof drainage channels. Flashing is normally of lead, copper, zinc or bitumen (Figure 3.6).

Wall

The gable end is the end wall of a building where the roof sits at right angles, above the wall. The hip end is the end wall of a building where the angled roof sits above all walls. The plinth is the mortar render to the base of the wall. Render is the mortar or pebbledash finish to the external wall. The damp-proof course (DPC) comprises a layer of impervious material (in plastic or slate, for example) integral to the wall, at approximately 150 millimetres above ground level, to prevent upward movement of ground water. DPCs can be injected where they are defective or absent.

Floors

The horizontal beams supporting the floorboards are the joists. These are normally of treated woodwork supported on joist hangers. Untreated woodwork (such as in Victorian housing) frequently has woodworm and is liable to rot due to lack of damp-proofing works, often going unnoticed. This subfloor structure needs to be ventilated by airbricks in the wall.

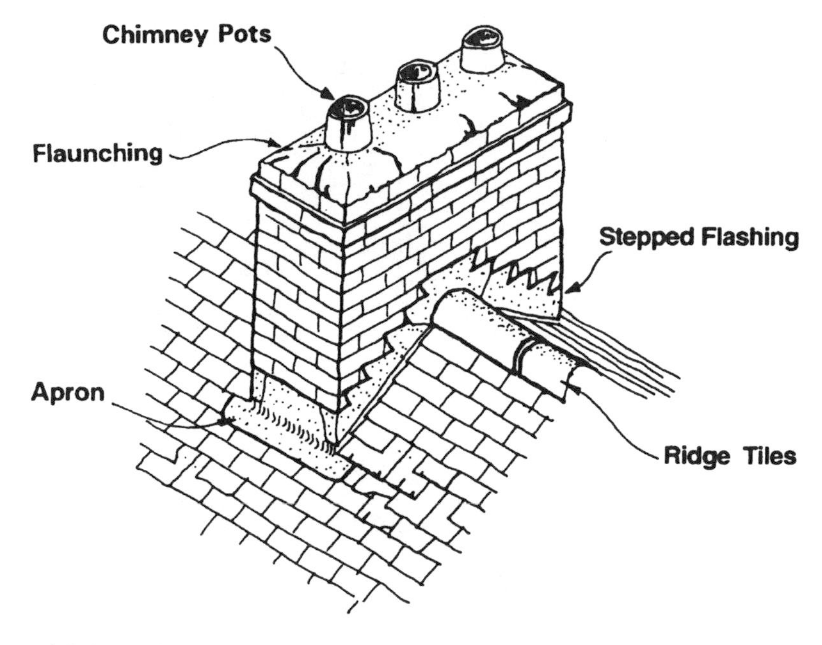

Figure 3.6 Parts of a chimney

Solid floors are concrete and require a damp-proof membrane, which is an impervious sheet of plastic, bitumen or similar, to prevent upward movement of moisture.

Stairs

The newel posts carry much of the weight of the staircase, supported also by the open and wall strings. The tread of the stair in effect forms part of the floor, whilst the risers add height with each step. The underlining of the staircase is the soffit, which often needs upgrading for HMO works. The handrail is horizontal, and the balusters, or balustrades vertical (Figure 3.7).

Plumbing and amenities

Amenities are served by water supply pipes, hot and/or cold, and drained by drainage pipes. The toilet pan is fed via the cistern, which has an overflow pipe. It has a water trap to prevent entry of both smell and vermin. Where the toilet is disused, it is useful to pour cooking oil in to prevent the water trap from evaporating (Figure 3.8).

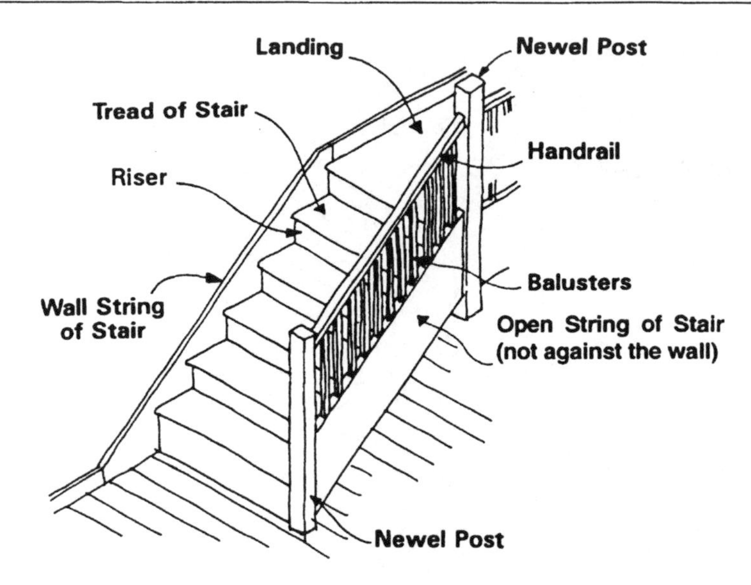

Figure 3.7 Parts of a staircase

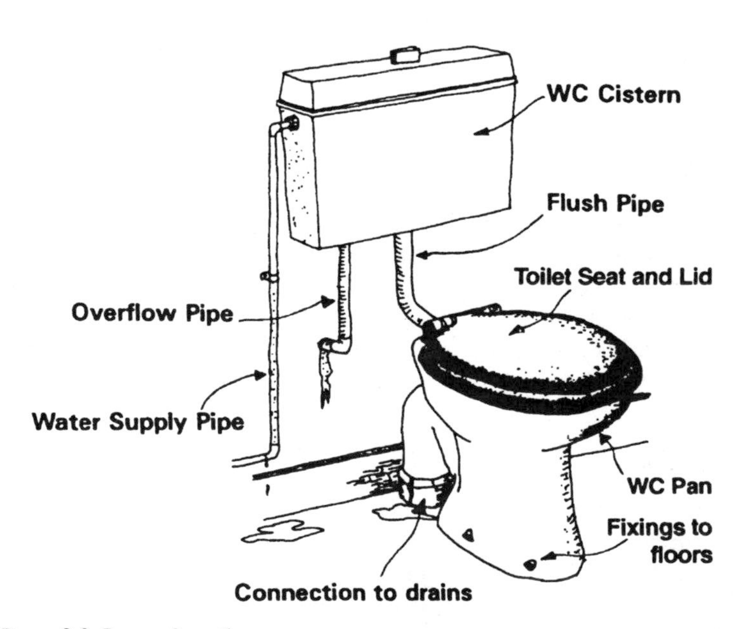

Figure 3.8 Parts of a toilet

Drainage

Drainage terminology can be split into both above and below ground drainage.

Above ground, the soil stack is the pipe taking foul water from toilet, sink and so on to the drain or sewer. It is sometimes ventilated above roof level to prevent water traps in the pipework being siphoned off. The hopperhead is a receptacle for serving several waste pipes receiving sudden large discharges of waste water, such as from a bath. The down-pipe is a vertical rainwater pipe (Figure 3.9).

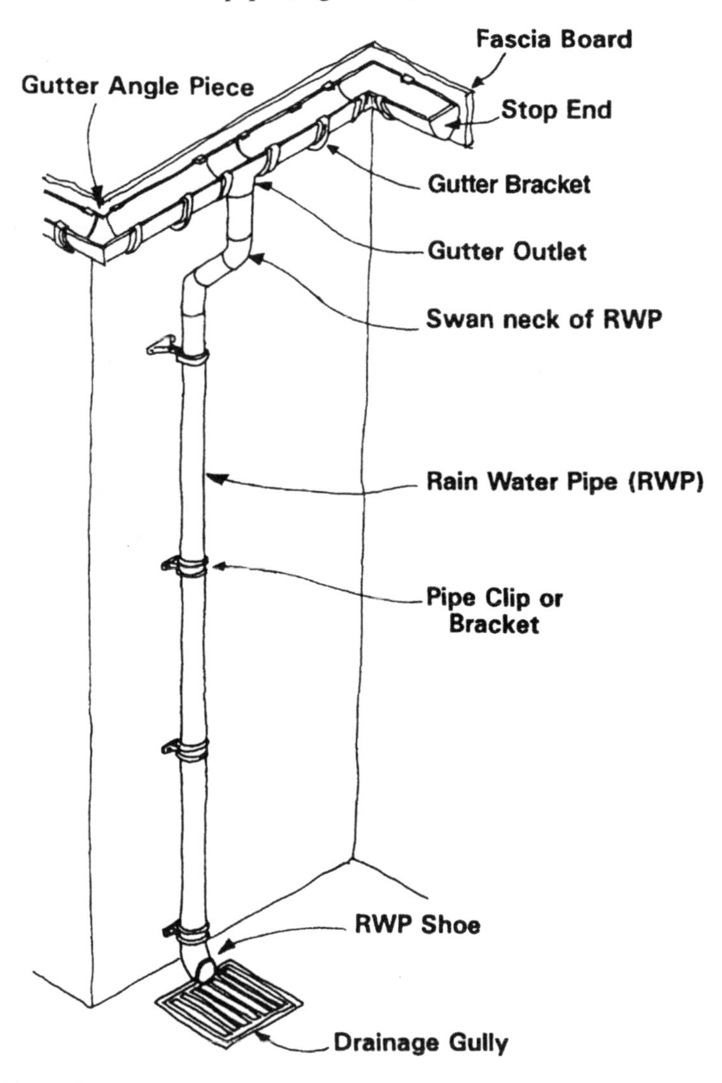

Figure 3.9 Parts of a gutter

Junctions of an underground drainage system are accessed via an inspection chamber (or manhole) cover. The inspection cover provides access to the system. In a trapped gully, the 'trap' retains water and, therefore, contains smells whilst still allowing proper drainage of waste water. The rodding eye provides an access point to unblock underground pipework.

Acknowledgement

The diagrams are reproduced with the kind permission of TRIS (now the Newcastle Tenants Federation), which has produced a number of other useful housing publications and which are available from TRIS, 1st Floor, 1 Pink Lane, Newcastle upon Tyne, NE1 5DW; tel.: 0191 232 1371.

Summary

- A good knowledge of key building terminology is essential as a basis in drafting specifications.
- There can be regional variations in design features and terminology.

3.2 Common housing defects

Outline

It is necessary to have a good understanding of the causes, nature and remedy of housing defects. There is not the scope here to discuss building defects in great detail as many other texts already do so. This section seeks to provide a brief introduction and to refer the reader to the Further reading section for more information.

General

There is a need to be accurate as the defect identification may form the basis of legal action. Sometimes the defect is extremely obvious. On other occasions, the defect(s) can be extremely complex and may require input from a specialist, such as a structural engineer or damp contractor. Table 3.1 provides a brief overview of common housing defects. See also Figures 3.10 and 3.11.

Table 3.1 Common housing defects – a brief overview

Defect	*Features*	*Remedy*
General disrepair	e.g. broken window pane, rotten door, slipped roof slate, defective electrics (e.g. Figure 4.3)	Identify and rectify individual features to prevent further deterioration
Condensation dampness	Usually to external wall, no definite edge, associated mould growth (e.g. Figure 2.5)	May be due to poor construction; increase ventilation (mechanical or natural), insulation and obtain advice on one's lifestyle
Rising damp	Usually 1 metre height, presence of salts, no mould	Repair or install a damp-proof course, renew the plasterwork to above the damaged area
Penetrating damp	Definite edge, caused by external disrepair, e.g. cracked guttering, missing tile, cracked render with corresponding damp patch (e.g. Figure 3.10)	Locate and rectify the source of the damp
Fungal attack in timber (dry rot)	Can cause serious damage that may extend beyond timber; develops in damp poorly ventilated areas with visual hyphae and orange fruiting bodies	Specialist contractor is required to remove affected timber and chemically treat surrounding areas as appropriate to specific fungus
Fungal attack in timber (wet rot)	Less pervasive than above, but can develop in damp conditions leading to decay parallel to the grain	Specialist contractor is required to remove affected timber and chemically treat surrounding areas as appropriate to specific fungus
Insect attack in timber	Major cause of timber decay, displaying flight holes and frass – most commonly caused by common furniture beetle (woodworm)	Specialist contractor is required to remove affected timber (if extensively damaged) and chemically treat surrounding areas as appropriate to specific insect attack
Stability	Cracking to exterior wall, e.g. at ground floor level or above window lintel (e.g. Figure 3.11)	Obtain structural engineer's or surveyor's report to identify the cause and carry out appropriate remedial action
Gas installation	Inadequate heating facility, symptoms of carbon monoxide poisoning	Gas appliances are legally required to be checked annually (enforced by the Health and Safety Executive)

Figure 3.10 Penetrating damp. For brief description of features, see Table 3.1

Some practical tips

In coming to a decision about the nature of a defect, it can be useful to ask the following questions:

- Are there obvious external repairs, such as missing slates or a leaning chimney? Cracks? Are all walls vertical and roofs (horizontally) level end to end?
- Tap plasterwork with knuckles to see if it is hollow (has it 'blown'?).
- Feel the floor with one's heels – does it sound dull? Is it springy? Is it level?
- Open and close windows and doors – do they fit? Do fitments work?
- Do taps work? Does the toilet flush?
- Does the house 'smell' damp? Is your assessment of dampness backed up by protimeter readings?
- Does your key or pen sink easily into woodwork?

Summary

- A good knowledge of basic housing defects is essential, with referral to specialist contractors such as engineers, electricians or gas installers as appropriate.

Figure 3.11 Severe structural instability. For a brief description of features and a remedy, see Table 3.1. In this situation, no early remedial works were taken (e.g. underpinning) to arrest the early stages of cracking, believing to be caused by the roots of nearby trees. Lack of early action led to the problem becoming worse, and the lintel dislodged, becoming dangerous. In this case, the instability became so severe that the property eventually had to be demolished

3.3 Surveying a house

Outline

This section overviews how to carry out a survey, the equipment required and the initial information needed on which to base future decision-making.

Carrying out a survey

The nature of the survey undertaken depends on the nature of the visit. A full or part survey may be required. The survey may be part of a wider house condition survey (see Section 3.8), a proactive or reactive case-based survey, and the nature of outcome required frequently dictates the survey type. Whilst a full house survey is normally preferable and information obtained useful for local authority records, it is in some cases impossible and provision needs to be made for this, with arrangements for re-visits as appropriate. The main purpose of the survey, and possibly the objective(s) of the person requesting the survey, need to be considered before embarking on it.

There are three key stages to this process, which are equally important:

- Over-viewing the case history (if available).
- Carrying out a detailed survey (collecting information).
- Written report on the findings of the survey (arranging and presenting the collected information).

Key rules in carrying out a survey

Several items of equipment should be taken on an inspection

- Inspection sheet or paper.
- Clipboard.
- Pen (a biro may not work if it rains).
- Tape measure/electronic tapemeasure (if visiting alone).
- Torch.
- Container for samples (if necessary).
- Binoculars.
- Damp meter.
- Camera.
- Ladder (subject to insurance).
- Identification.

To ensure entry, it is normally necessary to arrange a mutually convenient appointment and to allow enough time for the inspection. It may take a couple of hours to do thoroughly, including time for discussion with the occupier(s).

Individuals will adopt their own method of inspecting a house. The inspection should be systematic so that nothing is missed. In general, the same method should be taken each time. This may include a quick overall assessment of the house and discussion with the occupier, followed by a thorough external then internal survey of the premises. It is advisable to always start in the same area of each house, moving either back or forward, or right or left, on each floor, so that no room is missed, particularly in a larger property. This also ensures that references to right or left are construed accordingly and are clearly understood both by the surveyor and by the person(s) receiving later correspondence (such as, 'rising damp to bay window, ground floor front right').

References should always be made by room location rather than use to ensure accuracy. For example, 'GFFR room' rather than 'Bedroom 2' in case its use changes and causes confusion later on. The position of the right and left in the survey should be noted as 'references to right and left are taken from the road frontage facing the building', or however they were taken.

A sketch plan should indicate the premises layout, approximate sizes of rooms and common parts, and location of windows, heating facilities, boiler, vent fans, and so on, similar to that shown in Figure 3.12.

The survey must be accurate and precise. Measurements and numbers should be properly recorded, and the exact nature and extent of the defect should be noted. Vague phrases such as defective guttering or major dampness are not adequate, and the survey should report that the guttering to the front elevation is corroded 1 metre left from the downpipe, or that there is penetrating dampness, which has blown 2 square metres of plasterwork from the second floor rear addition room. The external inspection should then reveal, for example, a corresponding choked gutter at this position. The main purpose of the report is to diagnose defects and not the cause. The report can note the defect, then the specification can require remedial works, for example, 'trace the source and extent of dampness to the . . .'. Sometimes the cause is not immediately clear, if so, assumptions should not be made. For example, what initially appears to be subsidence may have been caused by a wartime bombing, an event that could only be known by talking to the tenants.

Each room should be looked at systematically and it is useful always to look at the structure in order; ceilings, walls, floor, door, window, etc. This will then ensure that all defects are seen and recorded, such as an inadequate fire door in an HMO, or an unsafe heating facility and other matters, such as management issues to common parts.

The inspection should be practical. It is wise to turn taps on and off,

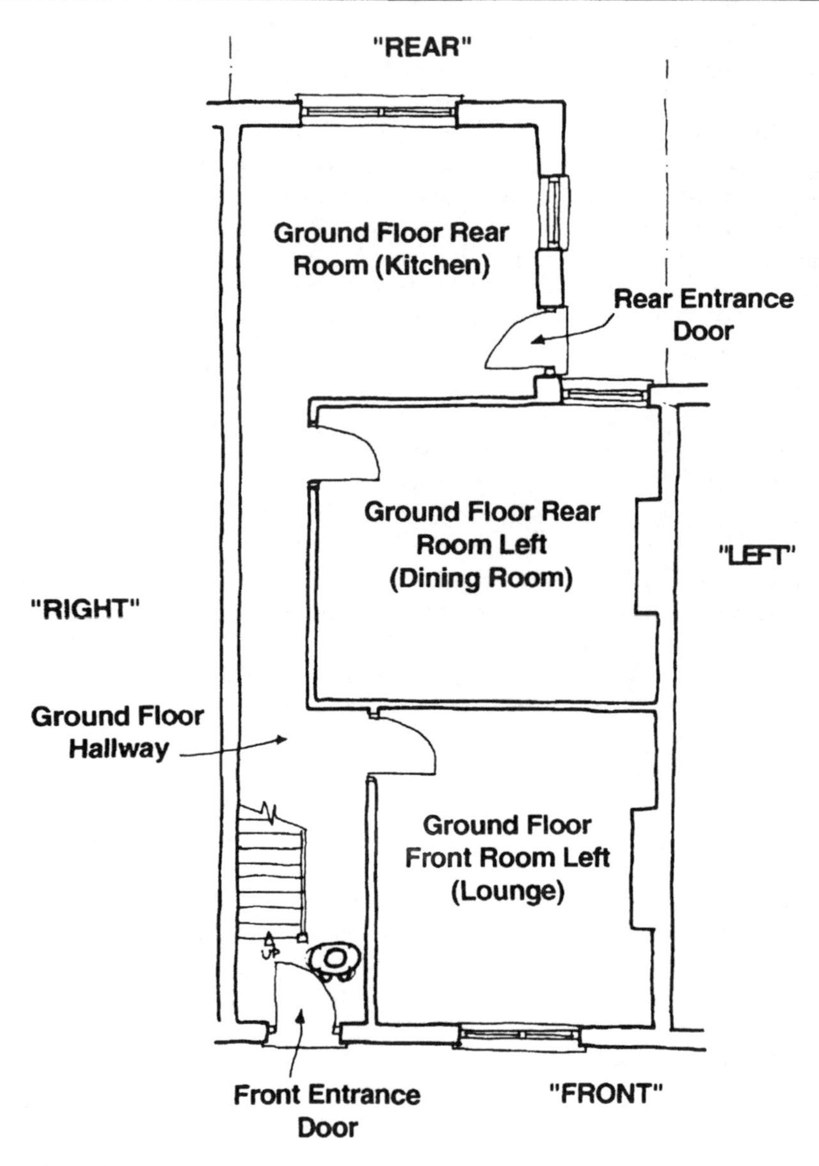

Figure 3.12 Ground floor plan of a typical terraced house

flush the toilet, open and close windows and doors. The occupier may bring items to attention and these should be examined objectively, such as an erratically operating light switch. If it works properly on inspection, the survey note should be that 'the tenant advised that . . .'. The survey report needs to be factual and accurate.

Equally, if something appears so defective that it might break, such as a rotten window, it is best not to touch it. The inspector should aim to make minimum damage when, say, using a damp meter or a key to see how rotten the woodwork is. Aim to cause minimum disruption to the furniture, carpets and so on, if at all. Health and safety is also important, for example, when using ladders or gaining access to a loft space.

Most local authorities have standard inspection forms (which serves as the survey and initial survey findings) and report procedures to follow. The Housing Act 1985 (as amended) requires that unfit houses should be reported to the council, and it is good practice to be in the habit of reporting on all issues so that anyone taking up the file is fully informed of the situation. In general, the report should be accurate and specific enough for others to follow as well as being an *aide-mémoire*. It should normally provide the following information:

- Sketch plan of dwelling, roughly to scale.
- Completed local authority survey sheet.
- Full details of occupier(s), type of tenancy, person to whom rent is paid, landlord, etc.
- Detail of nature, extent and location of defects.
- Report on premises.
- General idea of action to take, which may be based on risk.

The golden rule to consider when uncertain as to whether a defect is bad enough to warrant action is to ask, 'Would this bother me?' Use common sense.

Standard survey forms and/or checklists ensure that only relevant information is collected, and also that all relevant information is collected without the need to revisit. This might be in the format of an already designed survey form. They should be easy to adapt in case of legislative changes (such as in view of proposed changes to the fitness standard and HMO licensing; see Chapter 4). Several survey formats may be required to deal with different types of living accommodation, and the different types of information ultimately required, including a dwelling house, HMO, caravan or house boat. There should be space on the survey form to allow for additional and/or explanatory notes, sketches, photographs, risk assessment, and so on. An example survey form is illustrated in Section 3.9 (Figure 3.13).

Details taken at the time of the inspection must be accurate to provide a working description for inclusion in the report. If no access is possible, such as to a roof-space or to a covered floor, this should be noted so that the specification can accommodate for it. As much information as possible should be collated.

Drafting a report

A report should follow the survey. Part or all of this may already be incorporated into the survey form. In general, its arrangement should adopt the principles illustrated in Table 3.2.

Summary

- The first consideration is the purpose of the survey, which may inform how it is carried out.
- Survey details, including diagrams and notes need to be accurate and clear to understand.
- The survey should be practical in nature.
- Most local authorities have different survey forms relevant to different types of accommodation.
- A standard report format should be followed for consistency.

Table 3. 2 Drafting a housing report

Heading	*Example of commentary*
General	• Address of premises • Date of inspection • Tenants, tenancy type and date of commencement, contact details • Agent and landlord, address and contact details
Aim or purpose	e.g. 'To provide a general report and recommend remedial works' or 'the tenant had complained of dampness'
Description	e.g. 'This is a 1930s semi-detached house in a residential area. It is rendered externally and has a hipped roof. It has crittal windows. All floors are of suspended timber type. It is generally in poor repair'
Survey information	e.g. 'Roof. The original roof is tiled. The tiles are worn throughout. The soffit and fascia are rotting. The cast iron gutter and downpipes are corroding and loose'
Remedial works	Provide a brief overview of necessary remedial works, e.g. 'Roof. Renew roof tiles, soffit and fascia, guttering and downpipes'
Additional notes	e.g. 'One of the children suffers from asthma'
Conclusion	e.g. 'The dwelling is unfit because . . . and the most satisfactory course of action is . . .'
Recommendations	e.g. 'That a Housing Act 1985 (as amended) section 189 notice is served'
Action	e.g. 'Housing Act notice served on . . .'

3.4 Drafting a specification of works

Outline

The specification of works is the list of required works that will be sent, with or without a notice, to the person expected to carry out the works. It is essential, therefore, that the meaning, scope and extent of what is required is clear to all concerned. This section reviews the way in which specifications can be drafted.

Purpose of specifications

Specifications are written instructions describing ('specifying') specific items of work. They have two main functions:

- to provide instructions for the work content of specific operations; and
- to act as supporting/additional information to drawings.

They must be clearly and precisely written for use by architects, contractors, etc. for works in default and so on.

They are important to provide a clear, concise statement of exactly what is required, where and how. This is important in the case of notices and their compliance when requiring someone else to arrange for the works. Case law arising from appeals on notices served under the Housing Acts requires that the actual works, not merely their effects, should be specified and that care should be taken in their proper drafting (e.g. Moran 1997). The drafting of the notice needs to be precise and to provide sufficient information to enable the person served to obtain an estimate from a builder. Phrases such as 'as necessary', 'properly examine' and 'thoroughly overhaul' are normally considered satisfactory. The outcome of the extent of specification should be considered as a dwelling must be fit on completion of required works. It is all the more important is cases of works in default, where the local authority carry out the works required by notice where the person served has failed to do so in the time specified. This means that the local authority will then become the body responsible to ensure that works are carried out to required standards on a cost effective and professional basis.

Specifications are normally set into different sections:

- Preliminary specifications.
- General specifications.
- Standard details (e.g. drawings, specialist appendices, etc.).

Preliminary specifications

These relate to various aspects of the general organisation, information and administration of the required works. They frequently include:

- Site address.
- Scope of works.
- Form of building contract.
- Health and safety.
- Contractors' general obligations.

General specifications

These are generally combined in each item, but should include elements of material, workmanship and description of works as follows:

- Materials should be fully described making reference to British and other standards. The descriptions also include manufacturer's reference, size, colour, shape, etc. and to safeguard against defective materials.
- Workmanship includes the standards of preparation, fixing and finishing to required standards.
- Description of works is the full detail of required repair or improvement works.

Standard details

These are items that may be repeated in a typical contract and referred to by reference to the standard detail section rather than fully repeating the specification (e.g. for provision of a fire door) and can be appended.

Historically local authorities have prepared their own specifications in standard form for environmental health housing purposes to suit local circumstances. In practice, most local authorities now have computerised systems with purpose-written specifications provided with the software or using existing word-processed forms transferred onto the computer. This has the benefit of standardising the format, wording, efficiency and quality of work for notices and other documentation across the department. An example of a possible specification format is illustrated in Table 3.3.

Specifications

The three methods frequently used are:

- A direct order and brief specification/description of works. The cost is based on time, materials and additions for profits and overheads.

Table 3.3 `Example of a housing specification format

Heading	*Example of possible content*
Section A – Preliminary specifications	(i) Carry out all incidental and associated works necessary for the proper execution of the works described in this schedule provided the same may be reasonably inferred therefrom. (ii) etc.
Section B – General specifications	(1) Roof Strip off slates from the entire roof. Examine the roof timbers, replace all rotten timbers and strengthen the roof structure as necessary. Provide adequate ventilation, felt, batten and re-roof in accordance with current British Standards. Leave the whole sound and in a weather-tight condition. (2) Windows Take out the rotten or otherwise damaged frame and sashes to the ground floor front right window. Dismantle as necessary and cut out rotten or otherwise damaged woodwork. Fit new pre-primed woodwork to match existing, properly jointed. Reassemble and rehang. Leave sound, weatherproofed and in a proper working order on completion. And then following through and including the relevant subheadings and specifications as necessary, e.g. • Chimney stacks • Gutter and rainwater goods • Brickwork, stonework and rendering • Yard paving and external steps • External decoration • Internal plaster • Damp proofing works and penetrating damp • Doors • Subfloor ventilation • Fixtures, fittings and amenities • Services • Underground drainage • Floors and timbers
Section C – Appendices	Attach appropriate information to the specification such as 'Standard Details', or standard related other documentation, such as 'Automatic Fire Detection Systems', 'Emergency lighting', and so on, referring to this in the specification to prevent duplication of specific information

This is normally for small easily defined works and the most likely method to use on a notice or for grant purposes. They can be readily costed by reference to available pricing schedules.

- As above, but the works are taken from standard specifications that are completed as predetermined rates (i.e. 'Schedule of Rates'). This is normally issued where it is difficult to define certain works.

- The contract is carried out on a lump sum basis using a specification and Bill of Quantities and Drawings. This is used on large definable jobs with builders from a predetermined list. The builders competitively tender for the contract.

Summary

- Specifications need to be accurate, concise and precise.
- The format is likely to be determined by the local authority depending on whether the specification is for a notice, grant, works in default, large-scale contract, etc.
- The style selected should be consistent.

3.5 Key legal definitions for housing enforcement

Outline

An appeal can be upheld because an officer has mistakenly served the notice on the wrong person or organisation, or has made an error in a key legal definition. Such an apparently minor mistake can invalidate a notice, cause embarrassment and cost to a local authority and, most importantly, cause delay in a tenant getting necessary works done. To prevent this happening, a knowledge of basic definitions and where to find them is necessary. It is essential to have all relevant information to hand before instigating enforcement action as this could save substantial time in the long run. This section traces some key definitions and their application.

Understanding definitions

An overview of current legislation concerning living conditions is illustrated in Table 1.2 and should be read in conjunction with this section.

Definitions vary from Act to Act, some even within different parts of the same Act, and must be correctly used. For example, the statutory standard of fitness under the Housing Act 1985 (as amended) is not the same as that found under the Landlord and Tenant Act 1985. Similarly, case law seeking to define repair varies under the two statutes. Another example is with definition of overcrowding, where concepts are different under Parts X and XI of the Housing Act 1985 (as amended), and so on.

Whilst some legal terms are straightforward, others are far more complex and subject to continued debate through the courts. There are several recent examples of appeals brought for the interpretation and application of what constitutes a house in multiple occupation. Legal terminology can be difficult to understand. If uncertain of its meaning and potential application, it is useful to 'convert' it into simple terms, drawing out the key issues and supporting case law. An example of this is shown in Section 4.8, which illustrates the complexities of a single definition – a house in multiple occupation – and the case law that has built up over a period of years in attempts to apply it. It may also be necessary to refer to several definitions in seeking to clarify one, such an in the case of 'person having control', which is considered below.

The following definitions are only intended as a general guide and are not a specific interpretation of the law. Reference should be made to specific legislation and relevant case law according to the unique details of each case or situation, with further reference to a solicitor as necessary.

The following definitions are key to an understanding of environmental health housing law and practice.

Housing Act 1985 (as amended)

Building

Defined under section 207 to include the building containing a flat.

Dwelling house

Defined under section 207 to include a house or flat, and incorporates any yard, garden, outhouses and appurtenances belonging to it or used by it. This extends to flats, which are also dwelling houses, and notices may be served in respect of buildings containing flats. The definition does not apply to temporary or movable structures, which are separately defined and to which specific legislation applies. There is much case law to illustrate the definition, including Critchell v Lambeth B.C. (1957) 2 Q.B. 535, C.A., Quillotex v Ministry of Housing and Local Government (1966) 1 Q.B. 704, C.A. and Okereke v Brent L.B.C. (1967) 1 Q.B. 42, C.A.

Lessee

Defined under section 398 to include a statutory tenant of the premises and references to lease or to a person to whom premises are let should be construed accordingly.

House in multiple occupation

This definition is defined in Part IX, section 345 as a house that is occupied by persons who do not form a single household. This includes any part of a building that would be regarded as a house and building originally constructed or subsequently adapted for occupation by a single household. This definition, therefore, encompasses a single household's flat that comprises part of a larger building. A full breakdown of the definition is included in Section 4.8.

Occupying tenant

Section 207 states that an occupying tenant is not an owner-occupier, but may be a lessee, statutory tenant or specified occupier or licensee who is entitled to reside in the dwelling house because of their occupation.

Owner

Section 207 defines the owner as the person entitled to dispose of the premises, and includes a leaseholder with at least 3 years to run.

Owner-occupier

Defined under section 207 and may be an owner or lessee under the terms of the Leasehold Reform Act 1967 who occupies or is entitled to occupy the dwelling house.

Person having control

This is an important definition defined under section 207 as the person who receives the rent of the premises, or the person who would receive it if rent were paid, whether on his own account or as agent or trustee of another person. Rack rent means a rent that is not less than two-thirds of the full net annual value of the premises. In the case of a repair notice in respect of part of a building, the local authority should determine who ought to execute the repairs specified in the notice. It can be difficult to establish exactly who is the person having control in some cases, in which case advice should be sought from a solicitor.

Person managing

This definition is relevant to HMOs and is defined under section 398 as meaning the person who, being an owner or lessee of the premises, receives, directly or through an agent or trustee, rents or other payments from persons who are tenants or lodgers of parts of the premises. The definition also includes where those rents or payments are received through another person as agent or trustee to that other person.

Premises

Defined under section 207 to include a dwelling-house or part of a building, so this definition can be used for general residential premises, including those which fall outside of the pure dwelling-house definition. The definition under the Environmental Protection Act 1990 is different, and has a wider application potential.

Repair

Repair is not specifically defined under sections 189 and 190, but must relate to the actual physical condition of the building fabric that requires remedy. Repair generally means repair of existing and not improvement of existing, although the repair may constitute what could be deemed an 'improvement'. An example of this may be in the provision of a damp-proof course – required to prevent rising damp – that did not originally exist because it was not common building practice at the time of construction, but would now be an expected part of building construction, and could, therefore, be included on a repair notice. The proposed repair of each individual premises has to be considered and whether the proposed works would reasonably constitute a repair on grounds such as reasonable expense (relative to the value of the property), social issues surrounding the premises, current building technology and construction methods and its occupants. Substantial repairs might include one or more larger items. However, a repair notice under section 189 may incorporate an improvement such as provision of an internal toilet. Disrepair in itself is not the same as a house being 'unfit'. Much case law can be found under the Housing Act 1985 (as amended) and the Landlord and Tenant Act 1985.

Statutory standard of fitness

This key definition is defined under section 604 and first appeared in the Artisans and Labourers Dwellings Act 1868 ('unfit for human habitation'). It has been modified several times since, but is different in its statutory and common law interpretations. The current standard of statutory fitness is laid out in section 604, which is considered in more detail in Section 4.2, which also looks at proposals to introduce a fitness rating system. Reference should also be paid to Department of the Environment Circular 17/96 in determining the standard.

Environmental Protection Act 1990

These are defined under the sections 79–82 Statutory Nuisance provisions of the Act, and are further considered in Section 4.4.

Nuisance

Nuisance is not defined, but a statutory nuisance includes any premises in such a state as to be prejudicial to health or a nuisance. See Section 4.4 for further details.

Person responsible

This means, in relation to statutory nuisance, the person to whose act, default or sufferance the nuisance is attributable. Where the Environmental Protection Act is used instead of the Housing Act, this is generally considered similar to the 'person having control', so that the notice is served on the person receiving the rent, who would then coordinate the works.

Premises

This is defined to include any land and vessels, subject to some exemptions. This allows wider application than the Housing Act definition of premises.

Prejudicial to health

This is defined as being injurious or likely to cause injury to health. 'Health' is not defined. Further detail is included in Section 4.4.

Public Health Act 1936

Definitions for the filthy and verminous premises aspect of the Public Health Act are included in Section 4.12.

Summary

- Care must be taken in selecting legislation relevant to each case, and ensuring the correct use of legal terminology.

3.6 Introduction to landlord and tenant law

Outline

Action taken by local authorities in respect of housing conditions can generate some difficulties between landlord and tenant, particularly those in tenancies granted since 1988. This raises the need for officers to have a background knowledge in landlord and tenant law, its application and scope, and at what stage to refer to a housing advice officer for assistance and intervention for potential cases of harassment and illegal eviction. Local authorities need procedures in place to ensure close working between relevant departments and other organisations to assist tenants and where necessary landlords in advising on rights under legislation with a view to preventing, rather than aggravating, levels of homelessness.

This section briefly outlines private sector housing tenancies. It does not consider tenancies in social housing.

Nature of tenure

Tenure is indicative of the status of the person who occupies housing. The main forms are:

- Freehold (ownership).
- Leasehold.
- Tenancy (including a local authority tenant, private sector tenant, where different tenancies afford different tenant rights).
- Licence (including residents of hostels, family members of tenants).

Tenant

A tenant is someone renting property under a lease or tenancy in any housing sector. The term is sometimes interchangeable with 'lease', but tenancies tend to be issued for shorter-term arrangements. The main statutes in respect of private sector residential tenancies are the Rent Act 1977 and the Housing Act 1988. Public sector tenancies and licensees have protection under the Housing Act 1985: Part IV.

Lease

A lease is permission to occupy with exclusive possession on certain conditions, usually including rent. Permission to lease is granted by a person who is entitled to possession of the property (the landlord or lessor). Sometimes the landlord owns the freehold, or is a lessee under another lease (sublease). It may be for a fixed period or indefinitely until notice is granted bringing it to an end. Leases for more than 3 years must be granted using a deed. Leases tend to be more formal for longer-term arrangements, and tenancy for shorter-terms arrangements where no deed is issued.

Tenancy

A tenancy is permission to occupy a property with exclusive possession on certain conditions, including payment of rent. Permission is granted by the landlord, who may be the freeholder or another tenant. The tenancy may be for a fixed term or an indefinite period where notice is required to bring it to an end ('periodic tenancy'). There are many different tenancies, and arrangements based around the general notion of tenancy, including the following:

- Regulated tenancy – general term for a protected or statutory tenancy.
- Protected tenancy – private rented tenancy granted prior to 15 January 1989. Protected tenants retain the same status provided they remain with the same landlord. Fair rents can be registered with the rent officer, setting a maximum rent payable. There is security of tenure, so a landlord can only evict with grounds for possession.
- Statutory tenancy – tenancy created by statute when a protected tenancy has been brought to an end, for example where the protected tenant has died and someone is qualified to succeed to the tenancy such as a family member. The terms of the tenancy are the same, including security of tenure and the landlord can only get possession where there are grounds to do so. A landlord wishing to end such a tenancy must establish effective termination of the contractual tenancy or one of the statutory grounds for possession must apply, for example the premises required by former owner-occupier; holiday homes let out of season. There are also discretionary grounds for possession, including non-payment of rent; premises reasonably required by the landlord for his family. This means that in some cases a court has discretion as to whether to grant a possession order.
- Restricted Contract Occupier (under the Rent Act 1977) – includes furnished tenancies at low rent, tenancies that are not protected because the tenant shares some essential living accommodation with the landlord. There is very little security of tenure except under associated legislation, such as unlawful eviction under the Protection from Eviction Act 1977.

Rent Act 1977 and Housing Acts 1988 and 1996

Tenancies granted under the Rent Act 1977 have several advantages, including security of tenure and rent control through registration of a reasonable rent. From a tenant's perspective, less secure were the tenancies created by the Housing Act 1988, (see Section 2.3, Role of the private rented sector), where landlords were entitled to charge market rents with easier possession procedures at the expiry of the tenancy. This Act created the assured tenancies as follows:

- Assured tenancy – the tenant must occupy the dwelling as his/her only residence. It provides security of tenure. A landlord can only gain possession with a court order and must prove grounds for possession, such as nuisance or rent arrears. A judge would decide whether to issue a possession order on the basis of the landlord's written evidence, and a tenant does not have to leave until this is granted. Assured tenancies may be a fixed term (time limited) or periodic tenancy (running indefinitely and being continuously renewed for short periods, without the need for signing a new tenancy). It is the predominant tenancy for housing associations. The rent is generally set at a market level. One type is an assured shorthold tenancy.

- Assured shorthold tenancy – fixed term tenancies (for a minimum of 6 months) used mainly in the private rented sector, but sometimes by housing associations. The landlord must give the assured shorthold tenant at least 2 months' notice requiring possession of the property, the date of which must not be earlier than the end of the fixed term. Accelerated possession procedures can be used to regain possession when the tenancy has expired. This enables courts to make an order for possession without a hearing. It becomes a statutory periodic tenancy on expiry, and this tenancy can be ended by 2 months' notice at any time. The Housing Act 1996 makes all new tenancies assured shorthold, unless the landlord gives proper written notice otherwise. The effect of such tenancies is a lack of long-term security for many private sector tenants.

Other arrangements

Licence

This is permission to occupy a dwelling given to someone who is not a tenant or owner-occupier. By being granted a licence, the person is not a trespasser. Licensees include partners and family members. It is commonly associated with agreement between a landlord and occupier giving permission to occupy certain types of accommodation, including households with lodgers and shared housing, such as hostels. There are less legal rights than with tenants, no security of tenure and usually no right to exclusive possession. A landlord only needs to serve a notice to quit, usually giving 4 weeks' notice to require possession, but would need a court order if the licensee did not then move out.

Lodger

This is an individual living as part of the landlord's household. A lodger normally shares the household facilities, but with no right to exclusive possession to occupy any of the rooms, similar to licence status.

Squatter

This is a person occupying accommodation 'illegally' without permission of anyone entitled to give permission. A squatter is committing trespass. Court action for repossession by way of a temporary possession order can normally be obtained in a few days. A squatter commits a criminal offence on failing to leave the property within 24 hours of being served with a possession order. Mesne profits (i.e. money paid by illegal occupants of buildings or land to the person entitled to possession as compensation for the use of land and occupation of the property) can be claimed from a squatter as compensation for illegal occupation of the property.

Harassment and illegal eviction

Harassment and illegal eviction are commonly interrelated and can be difficult to prove because of the power differential between landlord and tenant. Harassment and illegal eviction are criminal offences and civil damages can be claimed. The Protection from Eviction Act 1977 (as amended by the Housing Act 1988) makes it an offence to:

- interfere with the peace or comfort of the tenant or anyone living with them; and
- persistently withdraw or withhold reasonable services needed to live in the premises.

This potentially has wide-ranging implications and may range from failing to pay a bill so that services are cut off, to antisocial activities against a tenant, to delaying repairs, to violent behaviour. Each case would need to be considered on its merits, and it would be more likely that the local authority would assist by, for example, arranging for reconnection of services or enforcing repairs through works in default, than by taking proceedings against a landlord for harassment and illegal eviction, or supporting a tenant in such a case, as this may provoke further problems between landlord and tenant, without getting the works done.

Eviction should normally be arranged through the courts, and all tenancies require written notification from the landlord seeking possession from the landlord, which accords to their tenancy, normally of 2 months. The tenant does not have to leave until then, and the landlord may have to seek a court order if the tenant does not leave at this stage. Different provisions apply to licensees, whose tenure is less secure, where notice would normally be equivalent to the licence period granted. Clearly, each case is unique.

Important note

The law relating to landlord and tenant can be extremely complex and the above is a guide only. Specialist advice should be sought from a housing advice officer or solicitor in individual cases.

Summary

- Most tenancies granted before the Housing Act 1988 are relatively secure.
- Tenancies granted after this Act are generally assured shorthold tenancies.
- Those without formal tenancy agreement are in a very insecure situation.
- Tenancy legislation is a very specialist area of housing law and advice from a solicitor may be required on the unique aspects of an individual tenancy and its status.

3.7 Basic private sector housing finance

Outline

Housing finance is a vital component of any private sector housing renewal programme. Government allowances and controls over capital and revenue expenditure have become increasingly tight in recent years, and are currently under review (DETR 1999). Management of expenditure has, therefore, become increasingly important both in terms of traditional housing finance, as well as seeking new and innovative resource expenditure. This section briefly overviews the importance of an accurate Housing Investment Programme (HIP), available funding opportunities and the need for auditing expenditure. It does not cover expenditure on housing association funding, which is overseen by the housing corporation.

General

Housing finance can largely be divided into two processes:

- Ongoing programmes of repair and improvement, including the HIP.
- Targeted regeneration programmes, such as the Single Regeneration Budget, or capital grants from the European Union.

Each of the above involves a substantial bidding process.

Housing Investment Programme (HIP)

The government makes annual housing capital allocations for local authority housing investment through the HIP. The programme is an annual process of preparing the local authority plans for housing expenditure across all housing tenures. The discretionary allocation process is closely regulated by guidance issued by the Department of the Environment, Transport and the Regions (DETR). Local authorities are required to provide details of their local housing strategy, successes in meeting this, and proposals for the future in comparison with other authorities in the region. They need to show how accurate their previous funding allocation has been in terms of outputs and all activities and expenditure must be fully justified. This process of 'bidding' for housing funding results in the Housing Capital Programme allocation.

The HIP bid incorporates expenditure plans for public and private sector stock, so requires cross-departmental input. Increasingly, policies have favoured the private sector and the DETR requires increasing information on this sector, including proposals in respect of:

- reducing unfitness and disrepair with associate grant aid;
- area-based action;
- initiatives in the private rented sector;
- strategies to tackle empty homes;
- care in the community; and
- domestic energy efficiency.

The HIP bid normally involves a presentation to the regional DETR office, followed by a formal submission. The presentation can involve the use of photographs and videos, which are useful to illustrate work undertaken or proposed. The regional DETR officials are then in a position to allocate their resources accordingly to the local authorities in their jurisdiction. They have considerable discretion in their funding allocation. Their decision is based upon existing data of needs indices, including the HIP returns, the annual survey of English Housing, the 5-yearly English House Condition Survey and the Census. The local authority's previous performance in delivering well-defined, local strategy within budget is also taken into consideration.

A major criticism of housing expenditure is that resource allocation is on an annual basis. This can make it very difficult for local authorities to undertake longer-term housing renewal projects in the absence of funding guarantees. Final HIP allocations are normally advised a few months prior to the new financial year, and expenditure proposals cannot be guaranteed until then.

Funding opportunities

Local authorities housing capital programmes are financed by a mixture of central government allocation and the local authority resources. The main function of capital programmes is for renovation and improvement of the authorities' own housing stock, grants to the private sector, disabled adaptations and grants to registered social landlords to support new build as well as rehabilitation schemes. This funding comes in four separate housing capital allocations (DETR 1999b):

- Housing Annual Capital Guideline (ACG).
- Allocation for Disabled Facilities Grants (DFG).
- Allocation for Private Sector Renewal Support Grant (PSRSG).
- Since 1997/8, a separate allocation for resources under the Capital Receipts Initiative (CRI).

Currently, Housing ACGs are a measure of local authorities' relative need to incur capital expenditure on housing in their area. This, together with other service's ACGs, form part of the local authority Basic Credit Approval (BCA), or permission to borrow to finance capital expenditure. The BCA is calculated by adding separate ACGs and then subtracting a proportion of capital receipts available as capital expenditure, known as the Receipts Taken Into Account (RTIA). This then provides an assessment of the local authority's ability to finance capital expenditure.

The CRI was more recently introduced to set-aside capital receipts from council house sales to be reinvested in housing. These resources are provided to local authorities through Supplementary Credit Approvals (SCAs) based on the level of receipts and on the relative need for housing capital investment, as measured by the General Needs Index. Release of these receipts needs to be closely monitored and spent effectively, and accountably, in terms of local housing strategies.

The vast majority of private sector housing renewal finance comes from the HIP allocation. Exchequer contribution varies according to grant type and area-based activity. For house renovation grants, allocation comprises 60 per cent of expenditure that is eligible for exchequer contribution, and local authorities must fund the remaining 40 per cent from their own resources. This is normally funded through the General Fund, Rate Support Grants and capital receipts. Local authorities, therefore, have to invest substantial finance in renewal, but failure to identify this resource would mean that they could not attract maximum potential exchequer subsidy.

Disabled facilities grants

Expenditure for private sector disabled facilities grants remains separate from other private sector renewal funding. Subsidy for disabled facilities grants is paid as Specified Capital Grant under HIP. Because these grants are mandatory and demand led, there is provision for Supplementary Credit Approvals.

Private Sector Renewal Support Grant (PSRSG)

This HIP allocation covers all discretionary private sector grant expenditure, including house renovation grants, renewal areas, group repair schemes, relocation grants, etc. These are paid as a Specific Grant, under the umbrella of Private Sector Renewal Support Grant (PSRSG). Because these grants are discretionary, funding is limited. If a local authority spent more than allocated, they would not receive subsidy above the amount granted, and would have to fund the full amount themselves. If they spent less, the DETR would reclaim the under-spend, which would affect future allocation.

The DETR makes payment for Specific Grant in ten monthly instalments following the first and second advance claims.

Single Regeneration Budget

SRB funding is becoming increasingly important in private sector housing renewal, particularly in terms of the wider economy and community participation in area-based schemes. It can be a useful component in attracting additional funding, or as an alternative to traditional HIP funding. Local authorities have to bid for this funding as and when it becomes available.

The DETR is increasingly looking to local authorities to demonstrate maximising their investment in private sector stock.

Auditing expenditure

Accountability for local authority expenditure has become increasingly important over recent years. In terms of HIP allocation and grant expenditure, this takes several forms. The priority is to spend properly the financial allocation in accordance with legal requirements. The management task is crucial.

In practice, most local authorities now have computer-based grant systems with several defined personnel involved in the grant process. This in itself can be a form of regulation and audit over expenditure. Good management is required to ensure that expenditure of allocation is accurate. Key personnel must ensure that works and grant payments, interim and final, are on schedule. This requires regular monitoring throughout the financial year. Such regular and thorough monitoring can also help detect fraud.

It is important that the local authority properly spend its allocation, as failure to do so can have a bearing on future allocations. It is useful for key personnel to meet regularly and review grant expenditure and commitment, so that any necessary revisions can be made at an early stage. Most computer packages can now carry out this task simply and quickly, as well as to collate statistics and financial information required by DETR returns.

Review of housing capital resource allocation – the single housing capital pot

The review of housing capital resource allocation is examining the needs indices used in allocating housing capital resources (DETR 1999). It is based on revised needs indices to take account of:

- Move to single allocation for local authority housing capital, the 'Single Pot'.
- Results from the 1996 English House Condition Survey stock condition data, which indicate regional differences in activity.
- Numbers of homeless in temporary accommodation as a measure of housing need.

Through this process, the government is currently seeking to provide local authorities with more freedom to determine their capital expenditure programmes, which proposed to move away from ring-fenced expenditure. The proposal is to merge CRI and PSRSG into the housing ACG, which is known as the 'Single Housing Capital Pot'. The change also seeks to target resources to authorities with greatest need, as well as reducing bureaucracy. One problem is that current allocation ratios in a single pot would reduce the private sector resource, so this is being reviewed.

Other possible developments in emerging housing capital allocation are also under consideration, with a view to improving needs indices. As the DETR (1999) notes, these aim to reflect current and emerging policy issues, priorities and developments and include:

- identification and treatment in the indices of low demand/unpopular housing;
- the emerging findings from other relevant Policy Action teams set to take forward the topics identified by the Social Exclusion Unit;

- the possible scope for the use of forward looking indices;
- the growing pressure on temporary accommodation in London and the South East; and
- the links between housing investment and wider regeneration issues.

Summary

- The HIP bid provides the framework for financial allocation, as well as for policy and strategy review, from the DETR.
- Traditional housing funding mechanisms are changing.
- Expenditure must be well managed, both to encourage future funding and to detect fraud in the system.
- Housing capital allocation is currently under review.

3.8 House condition surveys

Outline

The English House Condition Survey continues to show that there remain serious problems in conditions as well as housing distribution to meet supply and demand. As housing strategies become increasingly local in nature, there is a need, supported by government guidance, to maximise existing potential to meet need as well as to look to the future. This section traces the purpose and function of house condition surveys, before looking at their implementation and how to make best use of results obtained.

English House Condition Survey

The English House Condition Survey is a national overview of the nation's housing conditions and has been carried out every 5 years since 1967. The most recent was in 1996 (DETR 1998). It is the government's main source of information on conditions and helps determine current condition (including occupation and composition) and housing trends. The survey comprises four parts:

- Physical inspection of dwellings.
- Interviews with householders.
- Postal survey of local authorities and housing associations.
- Market value survey.

The main findings of the 1996 English House Condition Survey are illustrated in Table 3.4.

Already a picture starts to emerge of resulting government policy to

Table 3.4 Findings of the 1996 English House Condition Survey

- By the end of 1996 there were 20.4 million dwellings, with 19.7 million households
- Forty-five per cent of housing was more than 50 years old
- Eighty-one per cent of dwellings were houses. The remaining 19 per cent were flats, with about one-quarter of these being conversions
- There were 750,000 houses in multiple occupation, 62,000 of which were houses divided in to bedsits
- 450,000 had inadequate bedrooms, whilst more than 6.3 million households had surplus
- One per cent of housing stock (200,000 dwellings) lacked basic amenities, almost half of these being vacant, a reduction since the previous survey
- Sixty per cent of dwellings had double-glazing, a marked increase since the last survey
- Almost 80 per cent of dwellings had some form of defect
- 7.5 per cent of housing stock (1,552,000 dwellings) was unfit, the majority being pre-1919 and converted flats. Of these, 8.5 per cent could be made fit at £500 or less
- 19.3 per cent of unfitness is in the private rented sector, with 6 per cent in the owner-occupied sector
- Results indicated a significant correlation between household income and housing conditions.

Source: DETR (1998)

tackle poor housing conditions on a national scale, such as targeting of grants to lower income households, the availability of grants to the private rented sector where conditions are poorest, and the emergence of home repairs assistance to address low-cost renovation works. The survey also provides local authorities with a background to national housing conditions by which they can identify and develop programmes to deal with their own housing stock. Within this policy, local authorities also need to address local conditions through strategy, and the ways in which local conditions can be addressed is through appropriately targeted Local House Condition Surveys.

Local House Condition Surveys

Local authorities have a duty to consider housing conditions in their area on an annual basis and decide the action to take. This is an ongoing process, designed to build on existing information year on year. Local authorities need to identify housing conditions in their areas – the size, type, location, and condition and their suitability to meet local need, including for young single people, families, the elderly, the disabled and so on, both now and in the future. Wider planning issues, such as demographic change and local population growth, are also key in developing appropriate housing strategies. As Oxby (1999) illustrates, an imbalance between housing need and availability is demonstrated by a housing shortage as follows:

- Insufficient rented accommodation in the public and private sectors.
- High occupation densities with congestion and overcrowding.
- Constant pressure for rented accommodation.
- Multi-occupied houses with periods of occupation well beyond what might be considered reasonable.
- High land and house prices – consequently a depressed housing market which is difficult for first-time buyers to access.

This combination of factors has resulted in an increase in applicants applying to the local authority as homeless.

The nature and scale of change in housing need has led to local authorities having to review the way in which they deliver their housing services, with increased emphasis on making better use of existing stock. The way in which the suitability, availability and possible options for private sector stock can be assessed and can take many forms, with the most likely being a Local House Condition Survey, which can then inform the local housing strategy.

Initiating Local House Condition Surveys

While there is no specified way in which local authorities are required to undertake local surveys, the (then) Department of the Environment (DoE) issued guidance (1993) based on recommendations from the Building Research Establishment. The DoE categorised Local House Condition Surveys into three main types:

- Strategy development, for statistical and planning purposes.
- Action planning, for statistical and planning purposes.
- Implementation, for detailed specifications and costings.

Strategy development surveys

Strategy development surveys can be used to determine the:

- Nature and extent of problems.
- Appropriate levels of investment required.
- Priorities of action between tenures (such as public and private sector).
- Effectiveness of previous action and measuring schemes over time (such as group repair schemes).

The survey should cover both public and private sector housing stock but it is understood that the public housing stock may already have a programme of survey work. If so, this needs to be acknowledged in the HIP bid. The use of this type of survey is as an essential information source for HIP bid as well as for compliance with statutory requirements.

Action planning surveys

Action planning surveys are used where specific problems are already identified and they seek to focus on:

- costed options; and
- determining the relationship between housing conditions and household circumstances to determine appropriate solutions obtaining a detailed understanding of a specific subsection of stock, such as houses in multiple occupation (HMO), so that an active proactive programme can be pursued; or a geographic area known to have particular problems.

This type of survey is conducted within a limited area of, or a particular subsector of, stock. If area-based, it will focus primarily on one block or street. The intention is to develop and cost specific actions, but not necessarily for immediate implementation. The intention is for short-term planning in choosing between alternative options, setting priorities and establishing budgets. Reliable costs are required at block level rather than individual properties, including for group repair, common parts grants and so on. A sample survey can be used, but this needs to be large enough to include sufficient numbers of properties in each block.

Implementation survey

This type of survey is concerned with compiling a schedule of work for each dwelling and should include all dwellings identified as requiring intervention. This type of survey is intended to identify specific remedial actions for immediate implementation. The focus of this is on the individual dwelling so that the information obtained needs to be very reliable. Implementation surveys are generally confined to relatively small areas, such as individual estates or parts of estates. One hundred per cent coverage is necessary.

Local authorities may carry out separate surveys of public and private dwellings on the grounds that they want relatively detailed and non-comparable data for the two sectors. This is an acceptable approach. But local authorities need to ensure they can make comparisons between sectors to explain priorities and choices at the strategic level.

Specific detail on carrying out an individual survey is contained in Section 3.3.

Action following a survey

There is no point in carrying out a survey for the sake of it – the results need to be converted into effective policies and programmes for action. There are many uses of the survey information, which is increasingly held on databases and, therefore, can also be updated through on-going casework. The information obtained has many uses, including:

- Annual statutory requirement.
- To inform strategy.
- To identify areas or types of poor housing.
- To prioritise or assess risk work.
- To seek out types of accommodation, such as HMOs.
- To check local stock against national standards, in particular the English House Condition Survey.

Organisation and management of the survey

All local authorities already have a massive amount of information on housing conditions from previous surveys, local knowledge, casework, etc. in records in various departments, but this often seems to be uncoordinated. This is unfortunate since cross-referencing of information would help improve efficiency and reduce bureaucracy in collating housing information that already exists. Such information can, and should be, incorporated into on-going survey work, and increased use of suitable authority-wide computer databases may help improve this process. It is useful, as a geographical visual aid to survey work, and to maintain momentum, to have a borough-wide or ward-based map to identify reactive and proactive work previously undertaken in relation to private sector housing. This can be generated manually based on an officer's knowledge, or on a computer-generated GIS system where historic information is computer-archived. This might include information on private rented accommodation subject to environmental health enforcement activity, house renovation grants, group repair schemes and HMOs. This should help identify patterns and housing trends.

Instigating a new survey can be done in four main ways:

- In-house.
- Buy-in a commercial package.
- Contract out the job to consultants.
- Combination of the above.

There are pros and cons to each survey type, but management is the key. The benefit of the local authority participating in the survey includes the existing insight into the nature of the borough to obtain a broad overview, and also serves as an indirect form of staff training. With existing staffing levels already overstretched in some areas, this may not be viable, and so surveys tend to be contracted out. If the survey is contracted out, at least six organisations should be invited to tender. The DoE guidance on carrying out a house condition survey (DoE 1993) provides detailed advice on how to put out to tender and what exactly is required from the survey and the prices should be compared against the local authority's own estimate for carrying out the job. It is useful to consult other local authorities that have contracted out survey work.

One individual needs to take overall control of managing the survey, it is useful also to have a survey strategy group to develop new ideas and bring in experience from their departments. This also serves to share information between departments as a mutual learning process. The project must be effectively managed and should command appropriate resources, allowing sufficient time for day-to-day involvement.

District-wide strategic survey

It would not be feasible to survey all properties in a district, so the best representative sample should be obtained. The requirements of the survey must be carefully thought through so that resources are optimised and focused on certain areas or certain categories of building. A stratified sample (i.e. one that carries out a representative sample of different layers or types of properties) of 10 per cent total housing stock is generally recommended.

As basis for the HIP statement, local authorities need a position statement as to condition of stock in the borough to examine the extent and nature of problems to enable relative priorities between different sectors and priorities within sectors to be set.

The economies of scale of a single survey covering both public and private stock are evident. Separate surveys may be undertaken, but with a common core of questions. There are differences in how the data is to be used and stored. Generally speaking, private sector surveys are concerned with seeking an overall picture. Public sector surveys are more concerned with individual properties since their well-being is the sole responsibility of a local authority.

There are several advantages to a borough-wide strategic survey as follows:

- Provides general description of stock.
- Identifies immediate priorities for action.

- Enables information to be collated in an organised and coordinated manner.
- Enables resource implications to be identified to enable future planning.

Data handling

A district-wide survey would require an appropriate system of data handling due to the complexity and volume of incoming data obtained. For this reason, it is essential that any computer package is thoroughly researched before purchase or up-grading to ensure that it can provide, collate and facilitate the use of information obtained from the house condition survey. It is also essential to consider whether it is compatible with existing information on housing stock, including on-going caseworks and previous survey results. It is also important to select a system that has demonstrated success over a period of time in a similar authority.

Summary

- Local authorities are required to develop local housing strategies based on local housing conditions.
- The EHCS continues to illustrate high levels of unfitness.
- Surveys form the basis for making decisions about prioritising resources.
- Management follow-up from survey results is key in targeting resources wisely.

3.9 Housing practice and procedures

Outline

The purpose of this section is to provide a framework of environmental health housing law and practice, the 'bread and butter' of those dealing on a daily basis with private sector housing. It also seeks to assist those who need to know about such legislation on a more peripheral basis, such as some background information on fitness, or overcrowding. This section looks at what to do, when, where and how, when dealing with housing conditions and their enforcement. It provides a background to relevant legislation, serving notices, works in default and prosecution. Direct application of housing law and practice is detailed in Chapter 4.

Key legislation governing housing conditions

The key environmental health and housing legislation was summarised in Table 1.2. This section is concerned mainly with the procedures and practices associated with the following Acts and Circulars:

- Public Health Act 1936.
- Housing Act 1985 (as amended).
- Local Government and Housing Act 1989.
- Environmental Protection Act 1990.
- Housing Grants, Construction and Regeneration Act 1996.
- Housing Act 1996.
- DoE 17/96 – Private Sector Renewal: A Strategic Approach.

Whilst each Act is unique in its application, there are several principles that should be followed on a case-by-case basis to help provide a quality and uniform service that is regularly reviewed. Procedures should be well thought through and used consistently by all those involved in the housing process so that the client receives a 'seamless' service. Some of the following issues may sound obvious, but are frequently overlooked, and keeping to some simple rules can assist in providing a relevant and sensitive response in each unique case. Regardless of the relevant legislation, issues such as keeping case details, powers of entry, proper service of notices and, importantly, how notices are enforced in default are all key to the overall service each client receives, and how they might then perceive the local authority.

Keeping case details

Comprehensive details should be kept relating to each case. Increasingly such records are computerised on standard packages, but it is also useful to have back up paper-based case files. One file should be kept per case. These should incorporate all correspondence, including letters, inspection/survey forms (an example of part of one is illustrated in Figure 3.13), notices, telephone calls, risk assessments, copies of recorded delivery, details of how decisions were made, and so on. This is important in case of change of staff, ombudsman enquiry and quality assessment and a file should contain adequate details so that anyone looking at it would be aware of what has been happening, how decisions were made, who has done what and so on. It is useful to have a cover sheet at the front of each file summarising details of key actions and dates. A procedure checklist is useful to help ensure that all legally required, and other procedural actions, are properly followed. This also serves as an *aide-mémoire* and is particularly useful for new staff, or where different personnel are involved in one case.

Figures 3.13–16 as well as Tables 3.2 and 3.3 can be read together and illustrate how information obtained from an initial survey can be presented as part of the on-going case, and proceed in terms of serving a Housing Act notice requiring works.

Powers of entry

These differ slightly under various legislation, and entry must be sought under the relevant legislation to the case in hand, such as under the Housing Act 1985 (as amended) in respect of housing conditions, or the Public Health Act 1936 in respect of filthy and verminous conditions.

Generally speaking, the following procedure should be adopted:

- Informal approach, by at least one appointment letter.
- If no access allowed: hand-delivered letter (preferable in this case than recorded delivery) giving 24 hours' notice.
- If no access then allowed: application to the Magistrates' Court for a warrant to enter, setting out reasons why a warrant is needed and steps taken to date.
- Arrange to break in if necessary, with the police and/or other relevant personnel, such as the animal control officer, then leave secure.
- Warrant to be signed as executed, and original returned to the Magistrates' Court.

Service of notices

Who can serve notices?

Only officers with specific authority delegated under the Local Government Act 1972 can serve notices. Normally, the chief environmental health officer has power to authorise others to serve notices, normally by qualification and/or experience, and this may be outlined in the local authority's standing orders. It is usual practice for the task to be delegated to EHOs and housing technical officers. However, some local authorities still require the Chair of the housing committee to sign basic housing notices. Orders (e.g. Direction or Closing Orders) require the official seal of the council and special arrangements are normally made for their sealing and signing.

Notice procedure

A basic procedure can be adopted for all standard notices which helps ensure their proper service. Again, there may be specific procedures to follow under specific legislation with provisions for whom to serve. The

Main Occupier: MR A TENNANT	Date: 1 SEPTEMBER 2000
Address: 1 RENTING DRIVE LONDON Postcode: W1 Tel No: –	Inspector(s): MS. P.H. INSPECTOR

Owner: MR A ACKMAN	Agent: CATHY COMEHOME LETTINGS LTD
Address: 1 SPRUCE VILLA LONDON Postcode: W1 Tel No: –	Address: OCTAVIA TERRACE Postcode: W1 Tel No: –

Category	Fit	Unfit	Summary of Reasons
(a) Stability	✓		
(b) Repair		✓	ROOF & WINDOWS – RENEW
(c) Dampness	✓		
(d) 1. Lighting – Natural - Artificial	✓		
2. Heating	✓		
3. Ventilation	✓		
(e) Water Supply	✓		
(f) Preparation and cooking of food	✓		
1. Sink and Drainer	✓		
2. Hot and Cold Water	✓		
3. Layout	✓		
(g) WC	✓		
(h) 1. Bath/Shower	✓		
2. WHB	✓		
3. Hot and Cold Water	✓		
(i) Drainage – Foul – Surface	✓		

Is the property representative of adjacent properties? Better ☐ Same ☐ Worse ☑

Estimated age of property: 1910 Anticipated 'life' of property remaining: 50 YRS

Most satisfactory course of action: RENOVATION

Figure 3.13 Example of part of a housing survey record

ROOM INSPECTION SHEET NO

	Comments	Repair size (m^2)
Location FFFR		
Floor	SATISFACTORY	
Walls	"	
Ceiling	"	
Window(s) Sketch of design Size ratio to floor area	ROTTEN WOOD THROUGHOUT	ALL
Doors – int/ext	SATISFACTORY	
Heating	"	
Electrics	"	
Lighting – int/ext	"	
Ventilation – nat/artif	"	
Dampness (specify type)	PENETRATING DAMP DUE TO MISSING ROOF SLATES	$1m^2$
Fixtures and Fittings	SATISFACTORY	
Comments	NONE	

Figure 3.13 Continued

EXTERNAL

	Satis?	If unsatisfactory, state position and size (m^2)
Roof(s) (Type) SLATE	NO	WHOLE ROOF TO BE RENEWED – MISSING SLATES & SAGGING IN PLACES
Bays	YES	
Chimneys	"	
Firewalls	"	
Rainwater Pipes Fascia/Soffit etc	"	
Walls Stability	"	
Window and sills	NO	WOODEN CASEMENTS – ROTTED – ALL SIMILAR
External Doors	YES	
Ent Steps and thresholds	"	
DPC	"	
Drainage AG Drainage BG	"	
Yards/Paths Access/Fence	"	
Decoration Render/Paintwork Etc	"	
Outbuildings	"	
Sub-Floor Ventilation	"	
Comments	—	NONE

Figure 3.13 Continued

Local Government (Miscellaneous Provisions) Act 1976 provides for establishment of ownership details, etc.

One of the most important issues when serving notices is who to serve. If served on the wrong person or organisation, the notice will be invalid. Establishing ownership and occupation is, therefore, very important. Initially it is necessary to find a contact name, address and telephone number. This is then followed up by formally establishing ownership and other details using one or more of the following:

- Requisition for Information under Local Government (Miscellaneous Provisions) Act 1976 section 16.
- Land registry search.
- Specific legislation, e.g. the Housing (Management of Houses in Multiple Occupation) Regulations 1990, which provides for obtaining more direct and detailed information on occupation.

An example of a section 16 notice appears in Figure 3.14 and the reply format expected is shown in Figure 3.15.

Who is responsible?

Statutory provisions specify on whom notices should be served, for example 'person responsible', 'the person having control', and it is important to select the correct person or the notice will be invalid. These terms are defined in the relevant legislation, which also provides procedural rules governing the service of notices.

Details should be confirmed before action in relation to each case, but some general points apply as to whom to serve. The legal person or 'entity' to be served has to be recognisable in law as either:

- an individual (e.g. Mr Bloggs of . . . or Mr Bloggs trading under the name or style as . . .);
- a limited company (The Secretary, [Co.] Ltd);
- a group of people – who and why?;
- a corporation – an organisation which has a legal personality is called a corporation. Its existence is distinct from its members, such as a limited company. Public bodies such as local authorities are also corporations but operate to different rules than most others; or
- unincorporated associations, which include all other organisations and groups who join together for some purpose, but do not constitute a separate legal person. For example, clubs, voluntary organisations (except where 'incorporated') and partnerships. Partnerships are differentiated because they are associated with making a profit, so they have a distinct legal position.

Ref: HSG/1/2000

LOCAL GOVERNMENT (MISCELLANEOUS PROVISIONS) ACT 1976

REQUISITION FOR INFORMATION

To: THE SECRETARY, CATHY COME HOME LETTINGS LTD,
OCTAVIA TERRACE, LONDON, W1

1) IN PURSUANCE of its functions conferred by the provisions under section 189 of the HOUSING ACT 1985 (AS AMENDED), the CHADWICK BOROUGH Council requires you to state in writing the nature of your interest in the under-mentioned land and premises:

1 RENTING DRIVE, LONDON, W1

2) FURTHER, you are required to state in writing the name and address of any other person known to you as having an interest in the said land and premises and also to state whether that interest is a freeholder, mortgagee, lessee or otherwise.

3) The information requested should be furnished in the attached pro-forma within 14 days from the date of this notice.

Dated: (DATE OF SERVICE)

Signed [signature]

Designation ENVIRONMENTAL HEALTH OFFICER

Address of local authority

CHADWICK B.C.
TOWN HALL
LONDON
W1

Figure 3.14 Requisition for information pro-forma

In terms of serving the notice, the following general rules apply:

- If a corporation, the notice should be served on the Secretary (or Clerk) of the body at the registered or principal office.

Ref : HSG/1/2000

I set out below details of my interest in the following land and premises:

1 RENTING DRIVE, LONDON, W1

and also the interest of other persons therein. I believe that all the information given below is complete and correct.

1. My interest in the property is as — AGENT
2. The property is occupied by — MR A TENNANT
3. The occupier's interest is as (if as tenant state the term) — ASSURED SHORTHOLD (6 MONTHS)
4. The names and addresses of other interested persons:
   - Freeholder — MR R. ACKMAN
   - Mortgagee — N/A
   - Leaseholder — N/A
   - Person receiving rent — CATHY COMEHOME LETTINGS LTD
   - Any other interest — NO

Signed: [signature] Date: (TODAY'S DATE)

Name: FOR CATHY COMEHOME LETTINGS LTD

Address: OCTAVIA TERRACE, W3

Figure 3.15 Requisition for information pro-forma

- If an unincorporated organisation, the notice can be served on 'person(s)' (as defined under the Local Government (Miscellaneous Provisions) Act 1976, but it is questionable as to whether this is enforceable against the organisation.
- In the case of a partnership, a notice should be served on each partner by name in case of future criminal liability; in other associations, obtain as much information as possible about the organisa-

tion and legal advice as to who to serve may be necessary in uncertain cases.

- In the case of a sole trader, serve on Mr Bloggs trading under the name or style as Bloggs.
- Sometimes there is no one to serve notice on and it proves impossible to locate anyone to serve after extensive efforts. The Local Government (Miscellaneous Provisions) Act 1976 section 233 provides that if after 'reasonable enquiry' no one can be found, the document can be left with someone at the land or it can be left conspicuously displayed on some building or object at the land.

Drafting the notice

Requirements vary under each statute. The following is, therefore, an example only for serving notice under the Housing Act 1985 (as amended) section 189 (Figure 3.16), where the draft notice (for illustrative purposes only here) is based on the returned section 16 illustrated in Figure 3.15.

- Use the Housing (Prescribed Forms) (No. 2) Regulations 1990 and notice forms as cited.
- Serve a 'minded to' notice and respond to any communication arising.
- Specify a reasonable commencement time (greater than 21 days for the appeal period when the notice becomes operative, plus another 7 days is standard practice).
- Allow a reasonable time for completion (each local authority should have standard time scales for particular types of work).
- Specify the actual works to be done, not merely their effect. The specification must be precise enough for the person served to obtain estimates from a builder. Phrases such as 'as necessary', 'properly examine' and 'thoroughly overhaul' are generally acceptable (see also Section 3.4).
- (Consider serving a deferred action notice appropriate to the local housing strategy and review up to, or at, 2 years.)
- A copy should be served on anyone having interest in the dwelling including the owner, freeholder, mortgagee and lessee (legislation does not specify copying to a statutory tenant, although it is normally good practice to do so).
- Once the works are completed, the dwelling must be statutorily fit.

Reference HSG/1a/2000

HOUSING ACT 1985 (AS AMENDED) SECTION 189

NOTICE TO EXECUTE REPAIRS TO AN UNFIT DWELLING HOUSE

To THE SECRETARY, CATHY COMEHOME LETTINGS LTD, OCTAVIA TERRACE, LONDON, W3

1. You are the person having control of the dwelling house known as:
1 RENTING DRIVE, LONDON, W1

2. The CHADWICK BOROUGH Council are satisfied that the dwelling house is unfit for human habitation as described in Schedule 1 below.

3. Having regard to guidance given by the Secretary of State under Section 604A of the Housing Act 1985, the Council are satisfied that the most satisfactory course of action in respect of the dwelling house is the service of this repair notice.

4. In the opinion of the Council the works specified in Schedule 2 below will make the dwelling house fit for human habitation.

5. The Council require you to carry out the works and begin them not later than the 26 · NOVEMBER 2000 and to complete them within EIGHT (8) weeks of that date.

SCHEDULE 1

The dwelling house fails to meet the following requirements of Section 604 of the Housing Act 1985 (as amended):

(a) IT IS FREE FROM SERIOUS DISREPAIR

and by reasons of that failure, that dwelling house is not reasonably suitable for occupation.

SCHEDULE 2

Specification of works to be carried out

INSERT FIG 3.3

DATED this (DATE OF SERVICE)

Signed [illegible] Designation ENVIRONMENTAL HEALTH OFFICER

Cc
MR R ACKMAN, 1 SPRUCE VILLA, LONDON W1

Figure 3.16 Section 189 repair notice

Serving the notice

Local Government (Miscellaneous Provisions) Act 1972 sets out general rules for proper service by a local authority. This may be:

- delivery to the person;
- left at the proper address; or
- posted at the proper address (the last known address or specified address for postage).

Options for dealing with non-compliance

The purpose of the notice is formally to require, and to expect, works to be carried out. Service of notice does not always result in this and there are various ways in which a person served may be encouraged or forced into doing the work. Sometimes an informal approach, including continued site meetings, telephone calls and letters and/or the offer of discretionary grant, may go some way toward getting works done.

Where an informal approach fails to work, there are two main options. One is to carry out works in default, and the other is prosecution.

Works in default is extremely time-consuming and has to be precisely right, or the local authority might not get its money back. Each environmental health section should have a written policy for works in default that should be agreed with finance and legal services. It has to be watertight so that the local authority can fully recover its money plus expenses and interest, even where it involves putting a charge on the property.

Officers should discuss the possibility of prosecution, with full reasons, with senior officers and then the council solicitor to determine the best way forward in each individual case. Once again, the whole notice procedure must be accurate and watertight for a successful prosecution, highlighting the need for a good, thoroughly documented case file.

Summary

- Thorough case details need to be kept, both as a matter of good practice, and to provide a sound basis for enforcement action if this becomes necessary.
- Different legislation should be referred to directly for specific issues such as powers of entry, service of notice and so on.
- Local authorities need to have procedures in place to deal with prosecution and works in default in a prompt and professional manner.

Chapter 4

Environmental health and housing law and practice

Chapter 4 looks specifically at the environmental health housing law and its application. It considers the key concepts of housing fitness, nuisance and disrepair, before looking at options to deal with these. It covers issues in houses in multiple occupation, including definitions, management, amenities, general controls and means of escape in case of fire. It then turns to housing grants, considering their application and use. The chapter also explores alternative forms of living accommodation where conditions and controls are frequently tied up with land rights, before looking at where caravan site conditions are formalised and licensed. Some social aspects of housing are included, particularly basic public health issues found in filthy and verminous premises. The Home Energy Conservation Acts are considered. It provides an overview of powers to reconnect services in private rented premises where the landlord has defaulted in paying bills. It ends in looking at public assistance burials, where people have died with no one other than the local authority to arrange their funeral and to sort out their estate, sadly an increasingly common feature of society.

The chapter is presented as follows:

This chapter does not seek to provide full guidance on environmental health and housing law, as this can be found in *Clay's Handbook of Environmental Health* (Bassett 1999) and *Environmental Health Procedures* (Bassett 1998). It instead provides an overview and application to the legislation, which is illustrated with a series of case studies of how the theory relates, or otherwise, to the complexity of people's housing and lives.

4.1 Disrepair – civil and criminal remedy

Introduction and legislation

This section introduces civil and criminal remedies to tackle poor housing conditions. It shows the various remedies available under different Acts and how they might apply to particular conditions. Remedy for disrepair is available under the following legislation:

- Landlord and Tenant Act 1985 section 11.
- Defective Premises Act 1972 sections 1 and 4.
- Housing Act 1985 (as amended) sections 189 and 190.
- Environmental Protection Act 1990 section 80 statutory nuisance provisions.
- Other miscellaneous legislation, including the Building Act 1984 and various drainage legislation.

This section looks at civil remedy for disrepair, available across all housing tenures, and then at criminal remedy, which is applicable to the private rented sector and normally enforced by local authority environmental health departments.

Civil remedy for disrepair

Civil remedies are about private citizens' rights that are not criminal or political. In terms of remedy for disrepair, a tenant can sue their landlord for breach of contract, implied or explicit. Across both housing sectors (public and private), civil remedy can be sought through the following legislation:

- Landlord and Tenant Act 1985 section 11.
- Defective Premises Act 1972 sections 1 and 4.

Civil remedy under the Landlord and Tenant Act 1985 and the Defective Premises Act 1972 are now considered. The Environmental Protection Act 1990 nuisance provisions are considered in Section 4.4. Criminal remedy under the Housing Act 1985 (as amended) is discussed below.

Landlord and Tenant Act 1985 sections 8 and 11

The Landlord and Tenant Act 1985 section 8 contains implied terms as to fitness for human habitation, but is not widely applicable as it only applies to lettings at low rents. This section requires that the premises is fit on the day of letting and throughout the tenancy. The standard of fitness under this Act differs from the statutory standard of fitness under the Housing Act 1985 (as amended) section 604.

Section 11 is concerned with repairing obligations in short leases across all rented tenures. There is an implied covenant by the lessor to:

- keep the structure and exterior in reasonable repair;
- keep the gas, water and electricity services (including amenities, but excluding appliances) in repair and proper working order; and
- keep installations for space and water heating in repair and proper working order.

The disrepair must be such as to affect the lessee's enjoyment of the dwelling-house or common parts.

There are some provisions that apply in the application of section 11. The landlord must be in receipt of information from the tenant or a local authority repair notice to 'trigger' the liability to repair. There is an obligation to keep the whole building housing the premises in repair for leases granted after January 1989. The section only applies to tenancies and leases of less than 7 years. There is no obligation on the landlord to renovate after fire damage or similar. The tenants also have obligations – one implied covenant is to use the premises in a tenant like manner, without wilful or deliberate neglect. This obliges the tenant to keep proper care of the premises and carry out day-to-day jobs, such as mending a fuse or unblocking a sink. The Act allows for compensation claims for example for distress, damage to carpets and so on.

Defective Premises Act 1972 sections 1 and 4

The Defective Premises Act section 1 provides a duty to build dwellings properly:

- Applies only to building and repair work carried out after 1 January 1974.
- Requires use of proper materials and professional construction and to be fit for habitation on completion.
- Also applies to failure to carry out remedial work, or carrying it out poorly – may therefore apply to bad design, e.g. such as conversion to bedsits or flats, sound insulation or fire safety, as well as a landlord's programme of works to deal with an issue such as condensation that has been unsatisfactory.
- A landlord responsible for design and construction on a house let by him or her has a duty to take reasonable care to ensure that the house is free of defects likely to cause injury.

Section 4 provides a landlord's duty of care in virtue of obligation or right to repair premises demised and contains the following features:

- Applies to all tenancies (not just after 1974).
- The landlord has a duty of care to ensure residents and third parties are reasonably safe from personal injury (including disease and impairment of person's physical and mental condition) or damage by a 'relevant defect' (if the landlord is responsible for it under covenant, and only to parts let to the tenant).
- This applies if the landlord knows or ought to have known of the relevant defect (see Clarke v Taff Ely Borough Council (1984) 10 H.L.R. 44, Q.B. and Targett v Torfaen Borough Council (1992) 24 H.L.R. 164, so systematic inspection and maintenance is required of the landlord.

The duty of care only applies if the defect is a relevant defect, defined as a defect in the state of premises causing damage or injury because of a landlord's failure to carry out the obligation to repair or maintain, or an act or omission of landlord's liability had notice been given. Tenants and visitors can claim compensation for loss or injury resulting from failure to repair. See Figure 4.1 for action under tort.

Other possibilities

There are several other options available to the tenant. The Environmental Protection Act 1990 section 82 may be relevant for private action in respect of statutory nuisance (Figure 4.1) – this is covered in more detail in Section 4.4. Tenancy agreements may contain additional rights to repairs, known as repairing covenants. Access to free legal advice (Green Form) and expert reports, or lower rate legal advice from a housing solicitor may be available. Often claims are settled before

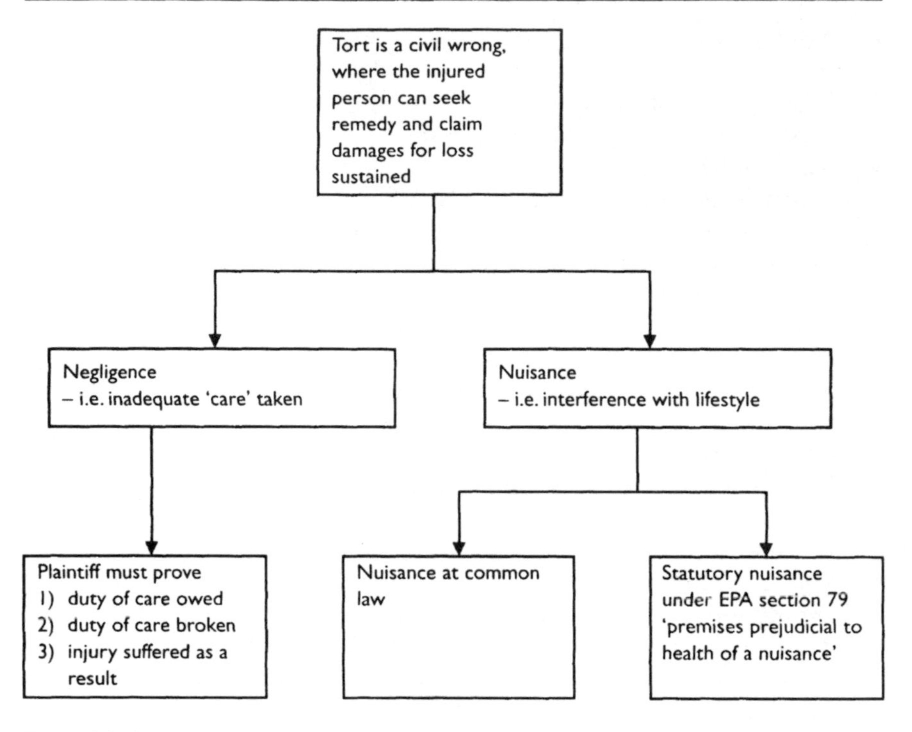

Figure 4.1 Action under tort

getting to court and it is generally up to the tenant to decide how to proceed.

Tenants can gain compensation under legislation via an ombudsman in cases of maladministration. Here there is no need for a lawyer, each circumstance is case based upon failure of a local authority to act properly. An ombudsman may investigate where a local authority tenant is aggrieved by failure to repair, or a local authority has failed to act properly for a private tenant in dealing with poor housing conditions. The investigation may find the local authority guilty of maladministration, delay or injustice, and this may provide a quicker remedy and more compensation to a tenant than remedy through the courts.

Criminal remedy for disrepair

It is normal practice for private tenants to secure their right to repair under statute enforced by local authority environmental health departments, although this does not preclude a civil case from running simultaneously.

Housing Act and disrepair

The Housing Act 1985 (as amended) provides remedy for tenants seeking repair and are enforced by environmental health departments on a daily basis. The Act provides for service of repair notices as a remedy both for disrepair as well as unfitness. This section concentrates on repair, whilst Section 4.2 looks at unfitness. The repair notice procedure is the same in each case, but with one major difference. Action to deal with disrepair is discretionary, whilst action for statutory unfitness (where repair is the most satisfactory course of action – see Section 4.3) is mandatory.

The Housing Act 1985 (as amended) provides for repair of dwellings as follows:

- Section 189 – repair notice in respect of an unfit house.
- Section 190 – repair notice in respect of a house in a state of disrepair but not unfit.

Definitions

Repair as a term is not defined. There is, therefore, a huge body of case law under the Housing Act 1985 (as amended) and the Landlord and Tenant Act 1985. This tends to revolve around the concept of making something good (which may result in improvement) rather than an improvement in itself. A fuller definition is provided in Section 3.5.

Repair notices – local authorities can require works of repair by notice under the Housing Act 1985 (as amended), sections 189 and 190. Section 190 notices are always confined to repair works, but section 189 notices can go beyond this, and require improvements. This is because of changes in housing law, such as compulsory improvement powers being superseded by a new fitness standard in 1989 which incorporated standard amenities, including an internal toilet, bath/shower, wash hand basin and hot water supply.

Application

Not all disrepair is sufficient to make a property statutorily unfit, but may cause tenants undue hardship. Local authorities should fully consider use of repair notices under the Housing Act 1985 (as amended) sections 190(1)(a) or (b) to tackle housing in disrepair before it becomes statutorily unfit.

In terms of repair of a dwelling-house, Lord Denning's precedent under Hillbank Properties v London Borough of Hackney (1978) Q.B. 998 summarises the importance of local authorities enforcing their discretionary repair powers as follows:

- Owners should be made to keep houses in proper repair, both to maintain the stock of houses and to ensure that protected tenants live in satisfactory conditions.
- The repairing power is important to compel owners of old houses to keep them in proper repair, where allowing them to fall into disrepair may be a means of evicting tenants.
- Where disrepair is serious (see Sections 4.2, on statutory fitness, and 4.3, on options for unfit dwelling-houses), it should be condemned, regardless of occupancy, through demolition, closure or purchase.
- If a house is worth repairing, it should be repaired, regardless whether occupied by a protected tenant or unprotected tenant.

Where a local authority is satisfied that a dwelling-house or HMO is in such a state of disrepair that, though not unfit for habitation either:

- 1(a) – substantial repairs are necessary to bring it to reasonable standard, with regard to its age, character and locality; or
- 1(b) – on representation by occupying tenant, conditions interfere materially with personal comfort of occupying tenant(s).

They have discretion to serve a repair notice. This is illustrated in Figure 4.2.

Additional provisions in serving section 190 notices

- There must be an occupying tenant, unless the dwelling is in a renewal area.
- A prescribed notice form is to be used.
- Notice to be served on the person having control or the person managing (if HMO). If the property is a flat or a flat in multiple occupation (FMO) and the reason for unfitness or disrepair is another part of the building, serve it on the person having control of that part.
- A minimum of 28 days should normally be allowed to start works – the notice becomes operative after 21 days if there is no appeal.
- Works required must not be of internal decorative repair (this is covered by the Landlord and Tenant Act 1985).
- Works are to be completed in a reasonable time as specified.
- Copy of notice is to be given to everyone with interest in the property, including the tenant.
- The notice is a local land charge.

Considerations in serving section 190 notices

- No requirement that 'repair' is the MSCA.

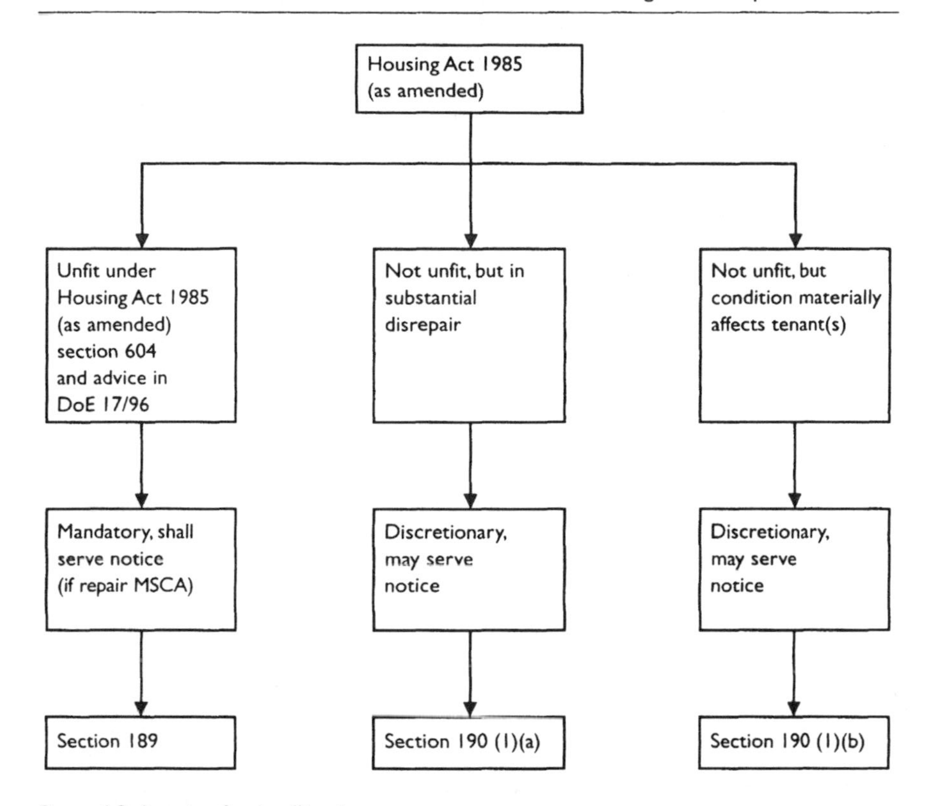

Figure 4.2 Repair of a dwelling-house

- Standard of 'repair' (slightly different under the Landlord and Tenant Act 1985) the test of 'reasonable expense' is relevant – the dwelling might otherwise fall into greater disrepair, which would be socially undesirable, and may point to deliberate neglect.
- 'Substantial disrepair' refers to one or more large items, or a combination of smaller ones.
- 'Materially affects' can be used where it would not be possible to use statutory nuisance legislation, which is discussed below.

Action following service of a repair notice under sections 189 or 190

- If there is an appeal to the County Court, the notice may be confirmed, varied or quashed (in which case the court may determine that local authority make alternate MSCA in case of unfitness, e.g. a closing order).

- If there is no appeal, the notice becomes operative after 21 days. If reasonable progress is not being made, the local authority can give 7 days' written notice, instigate works in default and recover costs.
- Intentional failure to comply with notices is a criminal offence. This could include, for example, not starting or completing work by required date. The local authority can prosecute the responsible party, even if concurrently instigating works in default.

For a flow chart on repair of house procedure, see Bassett (1998: FC107).

Summary

- Statutory notices can be served to deal with various stages of disrepair before the dwelling becomes statutorily unfit.
- Use of such notices remains discretionary, unlike notice to deal with unfitness, which are mandatory.
- Case law varies under different legislation, but is generally concerned with renovating what is already there and not provision of something new, which may be deemed an 'improvement'.

4.2 Statutory fitness for habitation

Introduction and legislation

The statutory standard of fitness is the key, national standard by which housing conditions are measured, but its definition and application have changed over time. The current standard is found under the Housing Act 1985 (as amended) section 604. A thorough understanding of the standard and ability to apply it is a fundamental environmental health function; it is the benchmark for housing assessment and triggers enforcement action.

Regular amendments to the definition have made it difficult to compare directly improvements in the housing condition. Although the Local Government and Housing Act 1989 sought to tackle unfitness in linking it to mandatory means tested grants, this had little overall impact on conditions, with the percentage of unfit dwellings reducing by approximately 1.3 per cent in 10 years from 1986 (Wilcox 1998). However, there has been progress in improvements made through provision of amenities, including an internal toilet, washing facilities and so on, and by 1991 most households had a bath or shower, although 100,000 still lacked an internal toilet (Conway 2000).

Unfitness is found across all tenures, but it tends to be associated with property age and conversion, with particularly high levels in the private rented sector. It comes as no surprise that unfitness is largely related to income, and occupied by the elderly and ethnic minorities (Leather and Morrison 1997). Whilst the English House Condition Survey illustrates continuing levels of unfitness, private sector grants have declined due to a lack of government funding.

A review of the fitness standard is currently underway and is briefly reviewed here.

Definition

There are several definitions of fitness, but this section is concerned with the statutory standard of fitness found under the Housing Act 1985 (as amended) section 604. This states that:

> A dwelling-house is fit for habitation for the purposes of the Act unless, in the opinion of the local housing authority, it fails to meet one or more of the requirements in a) to i) below, and by reason of that failure, is not suitable for occupation:
>
> (a) it is structurally stable
> (b) it is free from serious disrepair
> (c) it is free from dampness prejudicial to the health of the occupants (if any)
> (d) it has adequate provision for lighting, heating and ventilation
> (e) it has an adequate piped supply of wholesome water
> (f) there are satisfactory facilities in the dwelling-house for the preparation and cooking of food, including a sink with a satisfactory supply of hot and cold water
> (g) it has a suitable located water-closet for the exclusive use of the occupants (if any)
> (h) it has, for the exclusive use of the occupants (if any), a suitably located fixed bath or shower and wash hand basin, each of which is provided with satisfactory supply of hot and cold water; and
> (i) it has an effective system for the drainage of foul, waste and surface water.
>
> Whether or not a dwelling-house that is a flat satisfies the above requirements, it is unfit for human habitation if, in the opinion of the local authority, the building or part of the building outside the flat fails to meet one or more of the requirements below, and by reason of that failure, is not reasonably suitable for occupation:
>
> (a) the building or part is structurally stable
> (b) it is free from serious disrepair
> (c) it is free from dampness

(d) it has adequate provision for ventilation
(e) it has an effective system for the drainage of foul, waste and surface water.

Application

The statutory standard of fitness applies to all dwelling-houses across all tenures. It is a key standard in measuring housing conditions and triggers enforcement action, subject to the most satisfactory course of action as described in Section 4.3. An example of an unfit dwelling-house is illustrated in Figure 4.3.

Detailed advisory guidance on interpreting the fitness standard is given in advice provided in Annex A of Department of the Environment *Circular 17/96: Private Sector Renewal: A Strategic Approach* (1996). This circular considers each element of the fitness standard separately, providing a background, references and guidance notes on how the standard should be interpreted and applied. The guidance is strongly biased toward the health and safety element of the defect(s) and is based on the

Figure 4.3 Example of an unfit dwelling-house. This dwelling-house fails the Housing Act 1985 (as amended) section 604 statutory standard of fitness for several reasons, including being in serious disrepair (roof and render in particular) and not having a suitably located toilet (it is external here). In this instance, repair was considered the most satisfactory course of action (see Section 4.3) and a repair notice was served on the landlord under section 189.

physical characteristics of the dwelling-house, not on its occupation. Such guidance first appeared in Circular 6/90 in attempt to encourage uniformity in interpretation and application. An example of a house deemed statutorily unfit appears in Figure 4.3. Circular 17/96 also provides guidance on the most satisfactory course of action, which prescribes the procedure to take in respect of unfit dwelling-houses; this is dealt with in Section 4.3.

The definition applies to 'dwelling-houses', which includes any yard, garden, outhouses and appurtenances belonging to it or usually enjoyed with it (see the definition in Section 3.5). Whilst the condition of boundary walls, paths and yards can be taken into account, the poor condition of these alone would not normally be sufficient to render the house unfit, unless forming part of other defects (Oxby 1999). The definition also covers an HMO in its entirety.

Some elements to the standard are technical, such as provision of a wash hand basin or toilet, whilst others are more subjective, including the terms 'adequate' and 'suitable'. This has led some to suggest that despite government guidance, which sought to encourage objectivity, the standard remains essentially subjective, with no sense of degree of unfitness. Either a dwelling passes or fails the standard. There have also been criticisms that many poor health and housing issues, including, for example, fuel poverty brought about by inadequate insulation, or radon, are not included. The standard is not dynamic because it is based on enforcement rather than progress. A major criticism is that the current standard does not allow a rating of items, or the applying of an importance factor. For example, the lack of a wash hand basis currently ranks as highly as major condensation dampness. However, the standard is easily understood and is easy to apply to any given dwelling-house, although enforcement of the standard can sometimes be challenging. An example of its application is shown in Box 4.1.

Review of the fitness standard

In response to some of these concerns, the government is currently undertaking a review of the statutory standard of fitness. It proposed two ways in which the standards might be revised (DETR 1998). The first would add new criteria to the current standard, to incorporate energy efficiency, internal arrangement, fire safety and air quality. The second, and favoured, approach is to introduce a Housing Health and Safety Rating System. This will, first, address health and safety hazards not currently covered, and, second, move from concentrating on physical aspects, 'rating' the effect of the defects on the occupiers, so that degrees of unfitness can be differentiated (Toulcher 1998, Battersby and Ormandy 1999, Oxby 1999).

Box 4.1 Theory into practice – dealing with statutory unfitness and related issues

Sometimes it is clear from the outside of a dwelling-house that it is statutorily unfit. A neighbour was concerned for the elderly couple next door and had called in the environmental health section when the roof to their outside toilet was collapsing to see if the council could help. The EHO visited and looked from the outside, but no one would answer the door. The disrepair alone to the back elevation and addition warranted local authority intervention.

Returning to the office, the EHO set about obtaining information. There was no case file, but council tax officers confirmed the name of the owner-occupiers, and that they were elderly. The EHO drafted an initial letter saying that the local authority was surveying the area for targeting grants, and that their house had been identified in the survey, and an appointment was suggested. There was no reply to the suggested visit, so a second letter was sent. The EHO was surprised to receive a letter saying that the appointment was OK.

It was clear from the exterior that the dwelling would be unfit due to serious disrepair, but internally things were even worse than expected. It would turn out to be one of the most grossly unfit dwellings in the area. For starters, the EHO fell through the rotted floor behind the front door, before having to clamber over various articles along the way to the kitchen. The scale of works needed was massive relative to the size of the house. The outside toilet had practically collapsed and the new ceiling was bubble wrap, and a wooden contraption, with huge screws holding up the corroded toilet cistern. The tin bath was propped up in the corner. The dampness from this had penetrated into the kitchen, and the ceiling had partially collapsed. There was a cold tap to the butler sink and no hot water supply. There was no provision for heating. The original back door had rotted and finished about half way down, but this was apparently for ventilation and a good cat flap. And so it went on . . . .

But the required works were only part of the story. Beginning to gain the owner's confidence and get any works done required the utmost patience, perseverance and tolerance in a very difficult situation.

The house was in the woman's name, and her long-time partner had prevented her from carrying out any works because he believed he would do it all himself one day. This 'one day' had grown into decades, as the house had fallen first into disrepair, then to gross unfitness. His initial response to the council visiting was to paper the back bedroom with Mickey Mouse wallpaper and hold back the blown plasterwork with huge screws and wooden blocks. He also put some Old Spice aftershave on the paraffin heater to help freshen the place up. It was clear that it was going to be a long case. The elderly couple were offered rehousing, as formal closure was still a possibility, but the man did not want this and broke down into tears.

The local authority tried to offer all assistance it reasonably could, including an offer of temporary rehousing and a house renovation grant. Eventually a Housing Act notice had to be served on the owner because no progress was being made. The EHO pursued a grant for the owner, worried that otherwise the man would put his handy-skills to work, and she (as legal owner) seemed keen to go along with this. It was, however, obvious that even a maximum grant, with the owner's contribution, would be far short of covering all of the required works. The council, meanwhile, was able to secure temporary accommodation for the couple, and their cat, which had been an important issue.

Lengthy discussions were held with the builder, who had been involved in many complex grant cases, but nothing like this before. Discussions focused on the best way to go ahead with the work with the strict limit on funding. A programme was agreed, trying to take at least some of the owner's wishes into account, including painting the front door, to maintain some level of trust. Works were carried out to make the dwelling safe – including works to dislodged lintels, renewing windows, providing an inside bathroom and providing a new ceiling and completely replastering the kitchen and providing a sink, work surfaces and gas heater (to the living room) and hot water supply. The works took weeks, and many of the builders, as well as officers involved, suffered stress at the conditions and difficult progress with the works. All were glad when the agreed works were finished, and the dwelling was at least statutorily fit, if still needing some repairs.

The woman was most pleased with her new kitchen and bathroom. She had never had a bath in hot water at home before, and could not believe how nice it felt. She said that it was a relief to have a hot tap at the new sink, because she was finding that carrying boiling water from the stove had been getting a bit too much for her. It turned out in the end that the man was quite pleased too, because now that there was no longer a toilet outside, there was more space in the back yard to practise his ballroom dancing.

Guidance on the Housing Health and Safety Rating System is currently being developed. It seeks to:

- provide a new standard for measuring housing conditions related more directly to risk, and trigger an alternate approach to intervention;
- add substantially to factors to be included in determining housing conditions;
- assess a cumulative effect, rather than the mere presence or absence of an item; and
- enable a rating to be assessed by relevant scientific research related to health and housing.

Guidance has been prepared based on statistically based current knowledge and research and considers the potential for risk reduction. This system involves identifying faults, before deciding whether these faults contribute to hazard categories. The next stage is to assess the severity of each hazard, which will categorise each into a rating of likely harm. The resulting score provides a hazard profile for the dwelling, a relative measure indicating the severity of individual hazard, and whether that dwelling is better or worse than another, irrespective of occupancy. It is possible that rating bands will be developed to determine necessary actions (Battersby and Ormandy 1999).

The current fitness standard is under review and detailed guidance being prepared on a new approach to assessing housing conditions. Whilst some welcome the new proposals, others are of the opinion that it could prove unnecessarily complex to administer, are concerned that there has not been wide-scale debate on the issue, and favour maintaining the current standard. At the time of writing, it remains too early to determine how the proposed rating system will be used in legislation.

4.3 Options for unfit housing

Introduction and legislation

Local authorities have a duty to consider the 'most satisfactory course of action' (MSCA) for dealing with unfit premises. Decisions concerning every single-dwelling-house identified as statutorily unfit has to be reviewed in accordance with the provisions of Department of the Environment Circular 17/96 to determine the best way forward in each case.

Definitions

The MSCA is used when dwelling-houses have been determined as statutorily unfit to decide a course of action to deal with them on an individual or area basis.

Application

Local authorities are required to have regard to MSCA guidance in Department of the Environment Circular 17/96, Annex B. This provides a variety of considerations in determining a suitable way forward for unfit dwellings. The five options for remedy are found under the Housing Act 1985 (as amended) and are:

- Repair – section 189.
- Closure – section 264.

- Demolition – section 265.
- Clearance – section 289.
- Deferred Action – Housing Grants, Construction and Regeneration Act 1996 Sections 81–85.

These are dealt with below.

Considerations in the most satisfactory course of action

Local authorities have to make various considerations in coming to their decisions regarding the MSCA. These include:

- Possibility of charging for fitness enforcement.
- Making reasonable charge (up to £300) to recover expenses.
- Consideration of circumstances of person subject to enforcement action.
- Preformal enforcement procedures.
- Preformal enforcement procedures, including 'minded to' notice allowing right of representation, unless conditions are urgent.

Formal action

Formal action must be taken within the context of the local authority's private sector housing strategy. Decisions must involve consideration of those involved, and depending on the extent of unfitness area identified, this may be just the house, or the wider neighbourhood? The enforcement choice must be the MSCA, backed up by reasons for coming to decisions.

The initial assessment includes a Neighbourhood Renewal Assessment (NRA). The NRA method to decide MSCA and to explain the action chosen is based on the geographical scale of the unfitness. The purpose of the NRA is to consider relevant economic, social and environmental factors in equal weighting in the MSCA, using present and future costings (Net Present Values). The local authority determines a consistent range of costs, which includes the long-term consequences of action. Action on unfit premises needs to take into account the likely effect of that action on the neighbourhood.

The initial assessment involves:

- Considering enforcement options and resources available in the context of private sector housing strategy and the practicality of each option.
- Life expectancy of premises, relationship of premises to neighbouring premises and their condition.
- Proposals for the future of the area.

- Views, circumstances, wishes and proposals of owners and occupants.
- If rented, the landlord's management record.
- Effect of option on the community of the area.
- Effect of proposals in local environment and appearance of locality.
- Possibility of renovation grant.
- Suitability of premises for inclusion in group repair scheme.

Decision on the course of action and the notice served

A repair notice is the normal course of action for an unfit dwelling-house to bring the property back to statutory fitness. The example shown in Box 4.1 is of a severely unfit-dwelling-house, where issues such as the cost of works, wishes of the occupiers and the fact that it was mid-terrace were important as determining repair as the way forward. Similarly, repair was determined to be the MSCA for the vacant properties shown in Figures 5.4–6 due in part to the importance of the local authority's vacant property strategy.

A deferred action notice in effect puts action on hold for up to 2 years. The notice states why the dwelling-house is unfit, works necessary to make it fit and other possible courses of action. The local authority may still take other action, even before the 2-year review. This notice was introduced to assist in resource administration, flexibility, the occupiers' wishes and to enable advice and assistance to the person served. The decision to serve a deferred action notice is based on the circumstances and ability to carry out repairs, advice and support, the health and needs, the risk of premises, and funding, including grants.

The decision to serve a closing order is based on considerations such as the effect on neighbouring properties, its appearance and the local community, the potential for alternative uses, wider area initiatives and the availability of alternate housing. In serving a closing order, the local authority is in effect removing a unit of housing accommodation from local housing stock.

In considering the option to serve a demolition order, regard has to be had to the availability of local accommodation for re-housing, the prospective use of the cleared site, the local environment, the suitability of the area for residential accommodation and the impact on the neighbourhood. An example of a dwelling-house where demolition was determined as the MSCA due to severe structural instability is given in Figure 3.11.

Finally, the decision to declare a clearance area needs to be considered with regard to the context of neighbourhood, as well as the:

- concentration and proportion of unfits;

- density and street pattern;
- availability and suitability of alternate accommodation;
- adjoining land usage;
- listed status of protected and commercial buildings; and
- consultations, proposed future of site, including private investment potential.

For flow charts on procedures identified, see Bassett (1998: FC106–9). A summary of possible actions is shown in Figure 4.4.

Summary

- There are five MSCA options: repair, closure, demolition, deferred action and clearance area.
- In brief, MSCA requires full consideration of social, economic and environmental results of the proposed action.
- The extent of consideration is based around the geographical scale of the unfit housing conditions.

4.4 Statutory nuisance in housing

Introduction and legislation

While the Housing Acts would normally be applied in cases of poor conditions, there is also scope to use the tort of nuisance in enforcing housing standards. While this would not normally be a first course of action, this legislation holds several advantages over housing legislation, the main one being that it can apply to living accommodation that cannot be defined as a 'dwelling-house' – an example being a houseboat or a caravan. Additionally, in some cases higher standards can be required than just meeting, e.g. minimum fitness standards.

Definitions

The Environmental Protection Act 1990 provides for action in respect of statutory nuisance. In understanding the concept of statutory nuisance, it is first necessary to go back a step and trace the development of this tort.

What is a nuisance?

Most people would have their own understanding of the term 'nuisance', and what they might consider to be a nuisance, but applying the tort can be complex. A thorough understanding of the definition is required in terms of how it might apply to residential conditions.

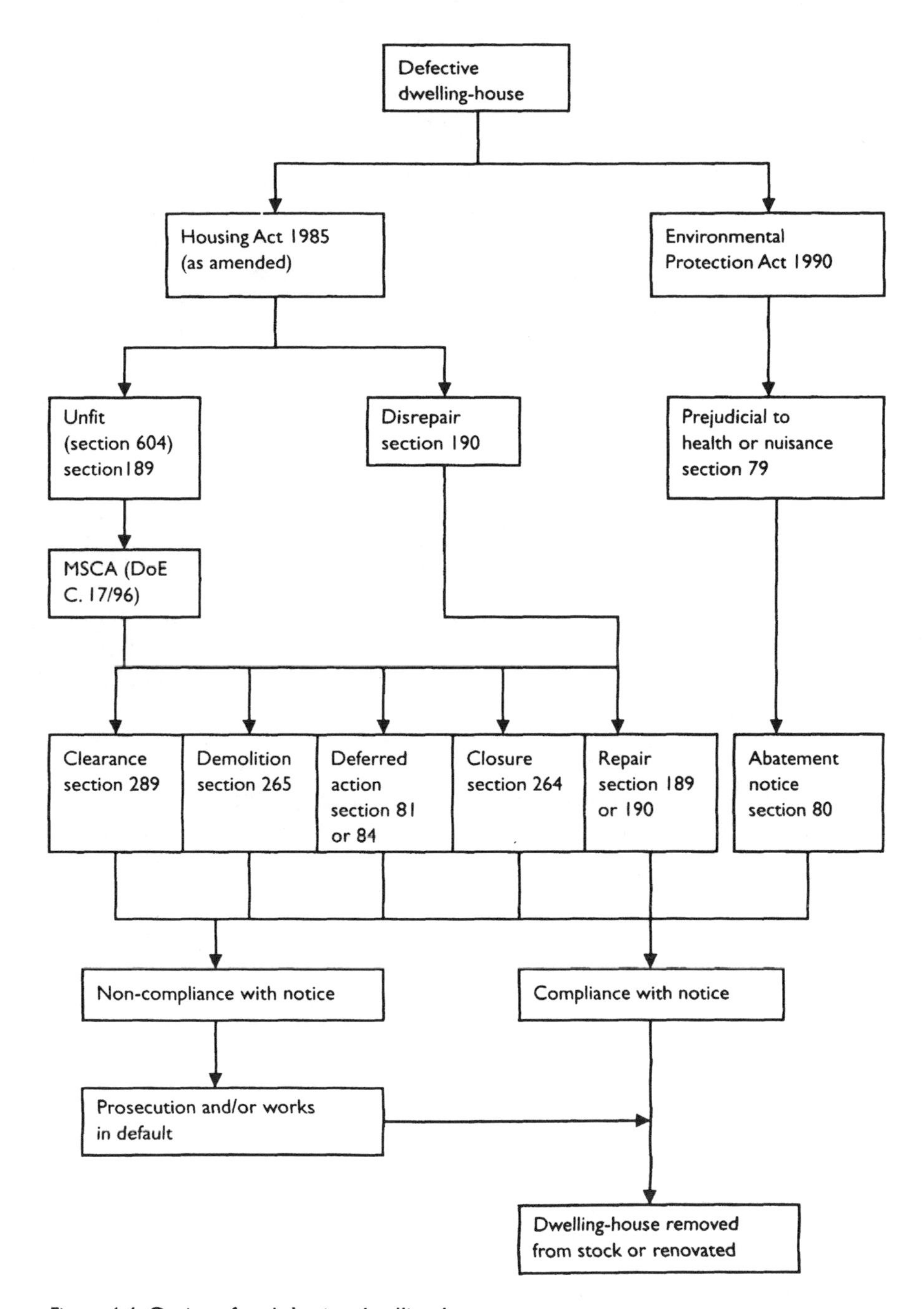

Figure 4.4 Options for defective dwelling-houses

At common law, nuisance can be split into public and private nuisance. A public nuisance is defined as an act that reasonably affects a class of Her Majesty's subjects, and there is no need for an interest in land affected. Conversely, a private nuisance is based upon unlawful interference with a person's use or enjoyment of their land or something causing an injury to that person. An example could include a leak of methane gas from a local landfill site. There is also scope for remedy where there is no actual physical harm, but nevertheless interference with personal comfort (Figure 4.1).

A statutory nuisance is a development of these basic concepts. The following key issues are relevant:

- Public nuisance is a criminal offence, but an individual can also institute civil proceedings and claim damages, an example being a contaminated water supply.
- Most public nuisances are enshrined in legislation and are listed under Environmental Protection Act section 79 (the defence of Best Practicable Means applies).
- The action of the nuisance must be substantial and unreasonable, either for a claim in common law or statutory nuisance. Case law illustrates that a malicious motive and location of the nuisance might be taken into account.

Statutory nuisance – the Environmental Protection Act 1990 sections 79–82

The Environmental Protection Act 1990 was introduced to restate the law defining statutory nuisance and to improve the summary procedures for dealing with them. This usefully means that previous legislation and case law can be followed. All local authorities have a duty to inspect their area from time to time to detect and act upon statutory nuisance. They also have a duty to inspect complaints of statutory nuisance. Where they do not, the Secretary of State has the power to act.

Defining statutory nuisance

The Environmental Protection Act 1990 clearly sets out local authority responsibilities in respect of inspecting for statutory nuisance as follows:

> Section 79: Statutory nuisances and inspections thereof
>
> 79 (1) Subject to subsections (2) to (6) below, the following matters constitute 'statutory nuisances' for the purposes of this part, that is to say –

(a) any premises in such a state as to be prejudicial to health or a nuisance;
(b) smoke emitted from premises so as to be prejudicial to health or a nuisance;
(c) fumes or gas emitted from premises so as to be prejudicial to health or a nuisance;
(d) any dust, steam, smell or other effluvia arising on industrial, trade or business premises and being prejudicial to health or a nuisance;
(e) any accumulation or deposit which is prejudicial to health or a nuisance;
(f) any animal kept in such a place or manner as to be prejudicial to health or a nuisance;
(g) noise emitted from premises so as to be prejudicial to health or a nuisance;
(h) any other matter declared by any enactment to be a statutory nuisance;

and it shall be the duty of every local authority to cause its area to be inspected from time to time to detect any statutory nuisances which ought to be dealt with under section 80 below and, where a complaint of a statutory nuisance is made to a person living within its area, to take such steps as are reasonably practicable to investigate the complaint.

Both 'nuisance' and 'prejudicial to health' can be a statutory nuisance, depending on the circumstances.

Nuisance

The key case to refer to is National Coal Board v Thorne BC (1976) 1 W.L.R. 543. This case illustrated how the spirit of the term 'nuisance' must be given its common law interpretation. This means that a nuisance cannot arise if it affects only the person or persons occupying the premises where the nuisance is said to have taken place. A nuisance, therefore, has to arise from one premises and affect another premises' occupants.

Prejudicial to health

This term is quite distinct. There is no need for the 'premises to premises' rule, so a statutory nuisance, using this basis, can arise in the same premises as the injured party. Key case law is Coventry City v Cartwright (1975) 1 W.L.R. 845, which also requires that the matter must be injurious or likely to cause injury to health. It is the effect of the defect, rather than the defect itself, which is important.

Section 79 (1) (a) is the relevant statutory nuisance provision for poor living accommodation, which encompasses 'any premises in such as state as to be prejudicial to health or a nuisance'. This section now requires further unravelling. What does it actually mean? An understanding can be obtained by breaking this legal provision into key areas:

- Any premises – defined to include all land, as well as including buildings and vessels.
- In such a state – where the statutory nuisance may result from one defect or accumulation of defects, the effect of which (not the defect itself or its source) must be likely to cause injury to health.
- Prejudicial to health – meaning means injurious or likely to cause injury to health. Health is not defined, but may be seen in terms of the World Health Organisation's definition (WHO 2000), namely, 'a state of complete physical, mental and social well being and not merely the absence of disease or infirmity'.
- Nuisance – the term should only be used where conditions potentially affect more than just those occupying the premises. In ensuring housing standards, it is likely that the 'premises in such a state as to be prejudicial to health', rather than the 'nuisance' clause will be used, as it is generally the conditions within the premises itself that would give rise to statutory nuisance.

Application

Abatement notices – the local authority role

The local authority role is clearly defined under legislation as including the following:

- Right to enter and inspect for statutory nuisance.
- Duty to act where statutory nuisance exists.
- Abatement notice served on the person responsible for nuisance, if structural works, serve on the 'owner' (i.e. the person receiving rent for repair, etc. works).
- Appeal within 21 days.
- A criminal offence if not complied with without reasonable excuse.
- The local authority can prosecute and/or instigate works in default.

Generally the local authority will be the prosecutor, but section 82 allows anyone aggrieved by the existence of a statutory nuisance to make a complaint to the magistrates court. Notice requirements under the Environmental Protection Act 1990 may follow the procedure below:

- An abatement notice to be served on the person responsible for the nuisance.
- Identification of the nuisance complained of.
- Requirements of the recipient – steps to abate or prevent occurrence of nuisance.
- Right of appeal.
- Time for compliance (depends on type of nuisance).
- Non-compliance with the abatement notice.
- Prosecution in a magistrates court for breach of the abatement notice.
- Taking action to abate the nuisance.
- Seeking an injunction in the High Court.

Advantages of the Environmental Protection Act 1990

The Housing Act 1985 (as amended) fitness standard only applies to 'dwelling-houses', which previously attracted a mandatory grant under the Local Government and Housing Act 1989 in cases of unfitness. The Environmental Protection Act 1990 was commonly used to prevent mandatory grant arising from Housing Act 1985 (as amended) notices. In addition, 'premises' can be widely interpreted and includes caravans, houseboats and self-builds. Nuisance legislation can be used in some cases to require works over and above requirements under other statutes. For options on notices, see Figure 4.4.

Accelerated procedures

Accelerated procedures for statutory nuisance are available under Building Act 1984 section 76 if unreasonable delay would result from the normal Environmental Protection Act 1990 procedure. Under the Building Act 1984 section 76, a local authority may serve notice specifying the defects and stating its intention to carry out remedial works within 9 days, unless the person served serves a counter notice in 7 days. The local authority cannot then act unless the person served fails to act within reasonable time, or is making slow progress.

If absolutely urgent, and 9 days would be too long to wait, for example in the case of a cracked and overflowing toilet waste pipe, the Environmental Protection Act 1990 can be used to require compliance in 48 hours. Such enforcement action should be backed up by sound works in default procedures.

For a flow chart on statutory nuisance procedure, see Bassett (1998: FC33).

Section 82 action by tenants

Private sector tenants tend to seek remedy via the Environmental Health Departments as per the procedure outlined above. In the case of public sector tenants, where local authorities do not enforce statutory provisions against themselves, there is remedy for private action under the Environmental Protection Act 1990 section 82. This section is used by other tenants, but less widely. These actions are criminal proceedings, started by laying of information. Successful action under section 82 may result in compensation and further litigation. The general procedure for section 82 action is as follows:

- Premises must be prejudicial to health.
- Tenant consults a solicitor.
- Solicitor obtains expert report.
- Person responsible is given 21 days' notice.
- Premises are still prejudicial to health after 21 days.
- Court proceedings are commenced.
- Contested hearing.
- Action by person responsible to remedy statutory nuisance.

Nuisance orders against local authorities

Existence of statutory nuisance has to be proved beyond reasonable doubt. The owner is liable to repair structural defects once statutory nuisance is proven. The court must make an order requiring works to abate the nuisance in a given time, and require sufficient works to prevent the recurrence of nuisance. This is based on the tenant's information and agreements reached, which may go beyond normal repairing obligations.

Following a successful nuisance order, penalties such as a fine may be imposed. It may offer tenants a quicker route to financial compensation than taking civil proceedings and sometimes the maximum possible is offered by the courts. Courts may also make an order for costs and legal aid. Based on expert witnesses, some local authority tenants have been able to secure substantial remedial works. This might include an order to pay costs for an independent EHO or similar, medical reports, photographs, legal costs. These might apply even if the tenant loses the case, but has good grounds for taking it.

Summary

- Use of the Housing Act 1985 (as amended) would always be priority for housing conditions, but the Environmental Protection Act may be more relevant in some cases.

- The Environmental Protection Act 1990 provides a legal remedy for conditions prejudicial to health, which may go over and above other housing act standards, such as fitness.
- It can be used to require urgent action.
- Nuisance legislation may be applied to non-traditional accommodation that can be defined as 'premises'.

4.5 Overcrowding

Introduction and legislation

Overcrowding legislation is quite draconian, based on standards from the 1930s which have still not been updated as they currently stand in the Housing Act 1985 (as amended) Part X sections 324–8, still using imperial measurements.

Definitions and application

Overcrowding in this context relates to dwellings, which are defined as 'a premises used or suitable for use as a separate dwelling'. Case law defines this as something in which all major life activities are carried out, for example sleeping, cooking and so on, where there is no sharing of another's living accommodation, such as a bathroom, toilet or kitchen (see Wright v Howell (1947) 92 S.J. 26; Curl v Angelo (1948) L.J.R. 1756; and Cole v Harris (1945) K.B. 474). With increased use of HMOs (e.g. bedsit-type accommodation), a single room may be deemed a 'dwelling'.

There are separate provisions for overcrowding in HMOs under Part XI sections 358–64 which are discretionary. There are also provisions relating to the number of facilities in HMOs – but these are 'over-occupation' rather than 'overcrowding' standards. Overcrowding is based on space availability, whilst over-occupation is based on the number of available amenities. Officers should have regard to all provisions when considering overcrowding and occupation standards.

A dwelling is overcrowded under the Housing Act 1985 (as amended) Part X if the number of people sleeping in it contravene either the room standard or the space standard, which are now defined.

Room standard

The 'room standard' (section 325) is contravened when the number of rooms and the number of people is such that two people of opposite sex, who are not living together as husband and wife, must sleep in the same room. For these purposes, no account is taken of children under the age of 10 and a room means a room of the type used in the locality either as a bedroom or a living room. 'Room available as sleeping accommodation' means a room normally used as a bedroom or living room. More usually, the standard is only contravened when people have no choice but to sleep in the same room.

Space standard

The space standard (section 326) is contravened when the number of people sleeping in the dwelling exceeds the permitted number (defined below) which has regard to the floor area and number of rooms available for sleeping accommodation. First the number of qualifying rooms must be counted and measured so a maximum number per room can be established. This was previously governed by the Housing Act (Overcrowding and Miscellaneous Forms) Regulations 1937, which, although repealed, may be helpful in the absence of further guidance, but could be challenged on appeal

The 'permitted number' is the smaller of the number of units (persons) calculated by the room or space standard (Tables 4.1 and 4.2). The equivalent number is the actual number of units (persons) accommodated. The 'space standard' is contravened when the number of people accommodated exceeds the permitted number. In calculating numbers allowed, the following should be taken into consideration:

- No account taken of a child under 1 year of age.
- A child between 1 and 10 years of age represents half a unit.
- A room must be a bedroom or living room of at least 50 square feet.
- Measurements should exclude areas where vertical height is less than 5 feet and non-available floor space, such as a chimney or fitted cupboard.

Offences

No offence is committed if an application is made to the local authority for rehousing and all persons must have been already been living there

Table 4.1 Overcrowding – floor area of people accommodated

Floor area (square feet)	*Number of persons*
≥ 110	2
> 90 but < 110	1.5
> 70 but < 90	1
> 50 but < 70	0.5
< 50	0

Source: Housing Act 1985 (as amended)

Table 4.2 Overcrowding – rooms available for sleeping accommodation

Number of rooms	*Number of persons*
1	2
2	3
5	5
4	7.5
> 5	2 for each room

Source: Housing Act 1985 (as amended)

when the child became 10 years of age, except where suitable alternative accommodation has been offered. It is possible for a non-family resident to move out at the occupier's request when the child reaches 10 years to prevent an overcrowding offence.

Overcrowding of houses in multiple occupation

Local authorities have discretion to control HMO overcrowding on a room-by-room basis, for example for hostels, hotels and so on, under the Housing Act 1985 (as amended) Part XI sections 358–64.

If a local authority considers that there are excessive numbers of people on the premises with regard to the room available or it is likely that an excessive number will be accommodated, they may serve an overcrowding notice on the occupier or the person managing the premises or both. The local authority must first serve 7 days' notice of intention and ensure that all people living on the premises are informed, so that representations can be made. Contravention is punishable by a Level 4 fine. A local authority can enter to inspect without prior warning.

A section 359 overcrowding notice must state either the number of persons to use each room as sleeping accommodation or that a room is not to be used as sleeping accommodation. Special room limits may also be set where people in a room are under the age specified in legislation. There is no requirement to use the standards in this section and some local authorities have adopted their own standards, particularly in view of high usage of HMO temporary accommodation in London. The notice must contain either the requirement not to permit excessive numbers to sleep on the premises (immediate effect) or the requirement not to admit new residents (reduction in numbers should be achieved through natural wastage). Clearly many local authorities would adopt for the latter to reduce potential homelessness.

Such notices can be used so that the person served would be fully aware of the legal requirements. Also, once such a notice has been served, a local authority has power to require information on the number of individuals, families/households, their names and rooms used. Failure to comply is an offence.

For flow charts on overcrowding procedures identified, see Bassett (1998: FC113, FC121).

Summary

- Overcrowding standards remain the same as standards introduced in 1937.
- Local authorities can set their own overcrowding standards for HMOs.
- Overcrowding is not the same as over-occupation, which is used solely for HMOs based on the relationship of amenities to occupiers.

4.6 Housing grants

Introduction and legislation

The concept of private sector housing grants was introduced in the post-war period under the Housing Act 1949. It was not, however, until after mass clearance programmes and subsequent municipalisation of housing stock that grants were seen to help stop the cycle of decay, so preserving housing such that it would not require early clearance. The Housing Act 1969 introduced General Improvement Areas (GIA) in an attempt to concentrate grants by area to tackle poor conditions. One result was that the grants were not being targeted to the poorest areas of working-class housing, resulting in problems such as gentrification and landlords exploiting the system (see Section 2.1). As a result, the Housing Act 1974 introduced Housing Action Areas (HAA), which sought to target grants to where they were most needed.

It was not until the Local Government and Housing Act 1989 that the grants system was completely overhauled. This Act introduced mandatory renovation grants (linked to fitness standard), houses in multiple occupation grants, common parts grants, minor works grants and disabled facilities grants. All grants were to be means tested, except for disabled facilities grants. GIAs and HAAs were superseded by Renewal Areas, which were much larger scale and sought to encourage new partnerships to obtain funding, with an emphasis on wider issues than just housing.

The Housing, Construction and Regeneration Act 1996 broke the link of unfitness and mandatory grants, and replaced Minor Works Grants with Home Repair Assistance, which had wider application. There was a major shift of emphasis from mandatory grants to discretionary grants. This new Act sought to streamline the system and target resources to people and properties most in need.

Definitions

These are included as necessary in the body of the text and figures in this section.

Application

The very nature of grants has changed substantially in recent years. Local authorities now need continually to review the priority they can give grants in a climate of decreasing resources. Part of this change involves increasingly working with others to provide other services, for example, with home improvement agencies, discussed below.

Local authorities have also had to take on board the overriding principles of Housing Grants, Construction and Regeneration Act 1996, with a shift in emphasis from state assistance to individual self-help. However, local authorities still have a duty to consider housing condition in their area and to determine strategy. Despite the overall move to discretionary grants and owner-occupiers now carrying the primary responsibility to the repair of their houses, all local authorities are expected to make some provision for discretionary grant aid. They would otherwise be seen as failing their responsibilities as housing enabler.

As a result, local authorities are encouraged to look at new options for renewal works, which will no longer rely on capital grants. These options include working with new and existing partners to offer:

- Advice.
- Assistance in obtaining builders.
- Agency services, e.g. Care and Repair Housing Associations.
- Charities.
- Other private sector solutions.

Housing Grants, Construction and Regeneration Act 1996

The Housing Grants, Construction and Regeneration Act 1996 provides discretionary assistance for:

- House Renovation Grant.
- Common Parts Grant.
- HMO Grant.
- Home Repair Assistance.
- Group Repair Schemes.
- Relocation Schemes.

Each grant is briefly considered, and a summary is included in Table 4.3. Considerations in grant application and approval are illustrated in Figure 4.5.

Table 4.3 Key criteria for discretionary grant eligibility

Type of grant	*Purpose*	*Residence required*	*Eligibility and restrictions*	*Maximum payable*	*Certificates and conditions*	*Test of resources*
House Renovation Grant	• To make dwelling fit • To put in reasonable repair • To comply with Housing Act sections 189, 190 or 352 notice • For home insulation • For heating • To provide satisfactory internal arrangement • Other works	3 years, except for landlord or if in the renewal area	• Owner-occupier – freeholder or leaseholder with 5 years remaining • Landlord – as above and let on residential basis • Tenant – where liable for works under lease or tenancy	Normally £20,000	Owner-occupiers, landlords or tenants; 5-year grant condition period	• Means test of owner-occupiers (including long leaseholders), or tenants • Landlord's assessment subject to local authority procedure
Common Parts Grant	As renovation grant, but applies to common parts Those not eligible for grant must pay their share	Not applicable	• Landlord (freeholder) • At least 75% 'occupying tenants' (if liable in lease) • Tenants and 'participating landlords'	Not applicable	• Landlord – 5-year grant condition period • Occupying tenant – certificate of future occupation does not apply	Dependent on who applies; may be as a house renovation grant, means test and landlord's assessment

HMO Grant	As renovation grant, and also to make HMO fit for numbers (section 352) by providing amenities	Not applicable	Available to HMO landlords	Normally £20,000	• Five-year grant condition • Repayment on breach of notice	Landlord's assessment subject to LA procedure
Home Repair Assistance	Grant or materials for small-scale repair, improvement or adaptation works Local authorities to determine the works they will grant aid	Not applicable, but 3 years where a non-standard dwelling, e.g. house-boat or mobile home	For owner-occupiers and tenants where: • over 18 years • dwelling is main residence • has owner's or tenant's interest • have duty or power to carry out works	£2,000, or £4,000 over 3 years	None	Must be in receipt of income-related benefits, e.g. income support, family credit, housing or council tax benefit except if aged over 60 years
Relocation Grant	For purchase of a new home arising from clearance action	Not applicable				As for renovation grant

Source: based on Housing Grants, Construction and Regeneration Act 1996

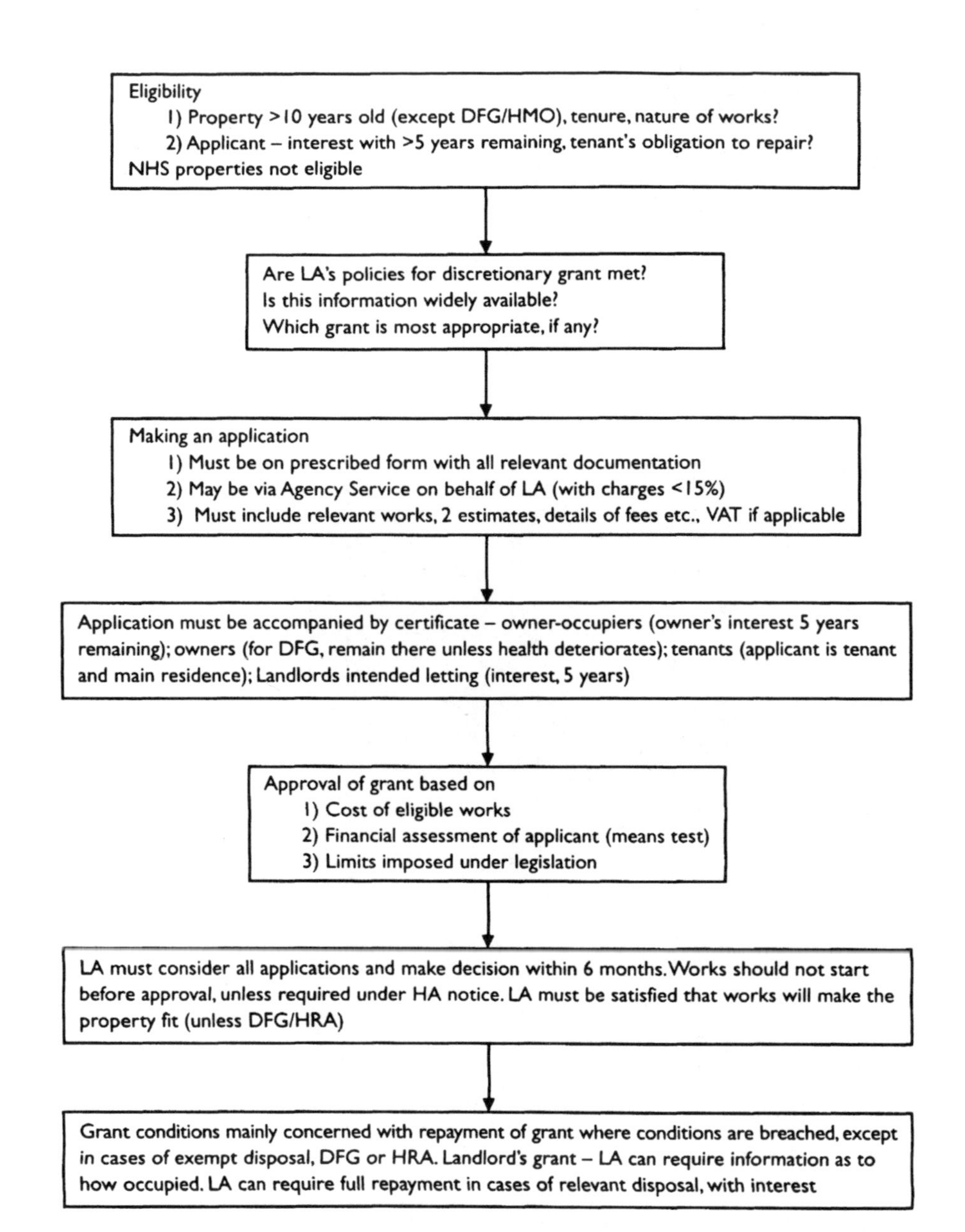

Figure 4.5 Considerations in grant application and approval

Renovation grants

The applicant(s) can be owner occupiers or tenants, who must sign a certificate of a 5-year grant condition to the effect that the house will be occupied as above. The purposes of a Renovation Grant are:

- to make the dwelling fit;
- to put it into reasonable repair;
- for energy efficiency (or HEES or HRA, see also Section 4.15);
- to provide satisfactory internal arrangement;
- for conversions; and
- for radon work.

Renovation Grants have a 3-year residency requirement prior to application, except where they are within a renewal area, or otherwise at the local authority discretion. They can also be used where housing act notices have been served, or to bring empty properties back into use.

For a flow chart on the house renovation procedure, see Bassett (1998: FC119).

Common parts grant

Common parts grants are for the improvement or repair of the common parts of buildings containing one or more flats. Eligible buildings include self-contained flats that are more than 10 years old, where three-quarters of the flats are occupied by occupying tenants (including leaseholders). The application should be made by more than three-quarters of occupying tenants, with a landlord as a participating tenant where appropriate.

Calculation or common parts grants can be complex. The costs are to be shared between those living in the building, determined as follows:

- Each applicant is means tested.
- The landlord's contribution is determined by the local authority.
- The local authority must ensure that everyone pays their share.

HMO grant

The definition of HMO for grant purposes excludes any part occupied by single household. Eligible works only include those required under the Housing Act 1985 (as amended) provisions, and not associated legislation that could, for example, be used for fire safety works. On completion, the HMO must be fit for human habitation and for the number of occupants under Housing Act 1985 (as amended). HMO grants can be awarded for conversion. The landlord has to provide a certificate in respect of the property and the intention to let or license it as a residence to people other than family members. The HMO grant is calculated at the discretion of the local authority.

Group repair schemes

The purpose of group repair schemes is to carry out external works to a group of properties at the same time using one contractor. This is more cost-effective and achieves a uniform renovation, so has an enhanced visual appeal. The types of works normally considered in group repair schemes includes renewal of chimney stacks, roof, gutters, rendering, walls, and so on depending on the original condition.

The works are organised and supervised by the local authority, which meets at least 50 per cent of the cost. All applicants are means tested, receiving a minimum of 50 per cent of the cost, with some receiving 100 per cent grant. A major objective is to help encourage owners to carry out external renovation works using a mixture of public and private finance. However, little research is carried out to follow-up whether this actually happens, and the extent to which households are able or willing to finance additional works remains questionable and needs thinking through. However, GRS have met with some interest, as shown by positive amendments in grant legislation from the 1989 to the 1996 Act, including providing greater flexibility and greater discretion to the local authority.

Buildings can be included in a group repair scheme if the whole or part of the exterior is not in reasonable repair, or if the structure is unstable and at least one primary building must be included. Different criteria apply depending on whether the scheme is within a renewal area. Before initiating a group repair scheme, a local authority needs to consider other options to ensure that it is the MSCA (see Section 4.3). They should take into account issues such as alternative uses for the site and other possibilities for dealing with the properties. Standard schemes do not need Secretary of State approval. Consultation is key throughout the project to ensure that all residents are kept fully involved. Buildings must be in reasonable repair on completion and have an expected life of at least 30 years.

Deveraux (1997) identifies the advantages of Group Repair over and above individual grants as including the following features:

- Strategic tool for improving the exterior of properties within and outside of renewal areas in small-scale group-based action, without the constraints of renewal areas.
- Provides for coordinated renovation.
- Creates the impetus for owners to invest in keeping the inside of

properties in good order and carrying out subsequent external maintenance.
- Encourages other investment in the area, such as environmental and energy-efficiency measures, so tackles fuel poverty, etc.
- Combines visual impact and financial assistance to make residents feel positive about their neighbourhood.
- Creates confidence in an area.
- Maximised resources due to economies of scale, so lowers costs to participants.

For flow chart on procedure, see Bassett (1998: FC110).

Relocation grant

Relocation grants seek to bridge the 'affordability gap' where people have to purchase a new home in the locality or another designated area as a result of clearance action. They aim to prevent the break up of communities and reduce the rehousing burden on the local authority.

For flow chart on procedure, see Bassett (1998: FC110).

Home repairs assistance

Home repairs assistance replaced and extended minor works grants under the Local Government and Housing Act 1989. Assistance (via grant or materials) is for small-scale repair or improvements where full-scale renovation is inappropriate. These grants have been extended to cover houseboats and mobile homes. Home repair assistance is a very flexible form of discretionary grant available to most people on benefits. Local authorities can target such assistance, for example, to elderly wishing to stay in their homes or to houseboat owners (see also Section 4.13). The grant can be used to promote thermal efficiency works, patch and mend in clearance areas, and so on. There is no prior residence requirement except for houseboats and mobile homes, where a 3-year period applies.

Non-eligibility for grants

There is non-eligibility for grants:

- If the property was built or converted less than 10 years before date of application (unless HMO grant).
- If the property is a second home or a holiday cottage.
- If (some cases) the applicant has not been in occupation as owner or a tenant for 3 years.

- If works are non-essential.
- If the applicant is a local authority tenant (except in the case of disabled facilities grants).
- If works are eligible for assistance under the Housing Defects Act 1984.

Grants and means testing

All grants to owners and tenants are now subject to a complex means test, with the exception of home repair assistance. The test normally applies to the applicant and their spouse or partner's income and savings. Where there are joint owners and tenants, the test applies to each of them individually. In the case of disabled facilities grants, the disabled person is assessed in a similar way, unless under 18 years of age, in which case the parents' means are taken into account. The test is to determine how much an applicant is deemed to be able to afford toward the cost of grant eligible works.

The DoE (1996) describes the process as follows. The test calculates the average weekly income and savings above £5,000. Savings below this level and certain state benefits are not included. This is considered against an assessment of basic needs, which form a range of allowances. If an applicant's resources are less than this assessment, they would not normally have to contribute toward the cost of works. Normally, anyone receiving income-related benefits will not have to make a contribution, but anyone else being assessed may raise the level above the threshold, so a contribution may be required. If an applicant's resources are more than the assessment, then a proportion of this is used to calculate the amount of loan, which the applicant is deemed able to afford. This 'affordable loan' is deducted from the cost of the eligible works, so the amount of grant approved would cover the remaining cost. Even if this amount is assessed as zero, this should be formally approved in case of later application. This process is summarised in Figure 4.6.

Landlord's grants are not means tested and local authorities have discretion on how they determine grants to landlords. Local authorities take issues into account such as the current rent level, expected capital and rental increase in the property arising from the works, the contribution of works to wider objectives, the nature of the works and the applicant's management record. This enables local authorities to prioritise a grant toward, for example, providing a means of escape in case of fire in HMOs, where works can sometimes reduce the capital value, or toward vacant dwellings as part of a wider strategy of renewal.

Home repair assistance is not means tested as such, but the applicant must be in receipt of an income-related benefit. This might include income support, an income-based job seekers allowance, family credit,

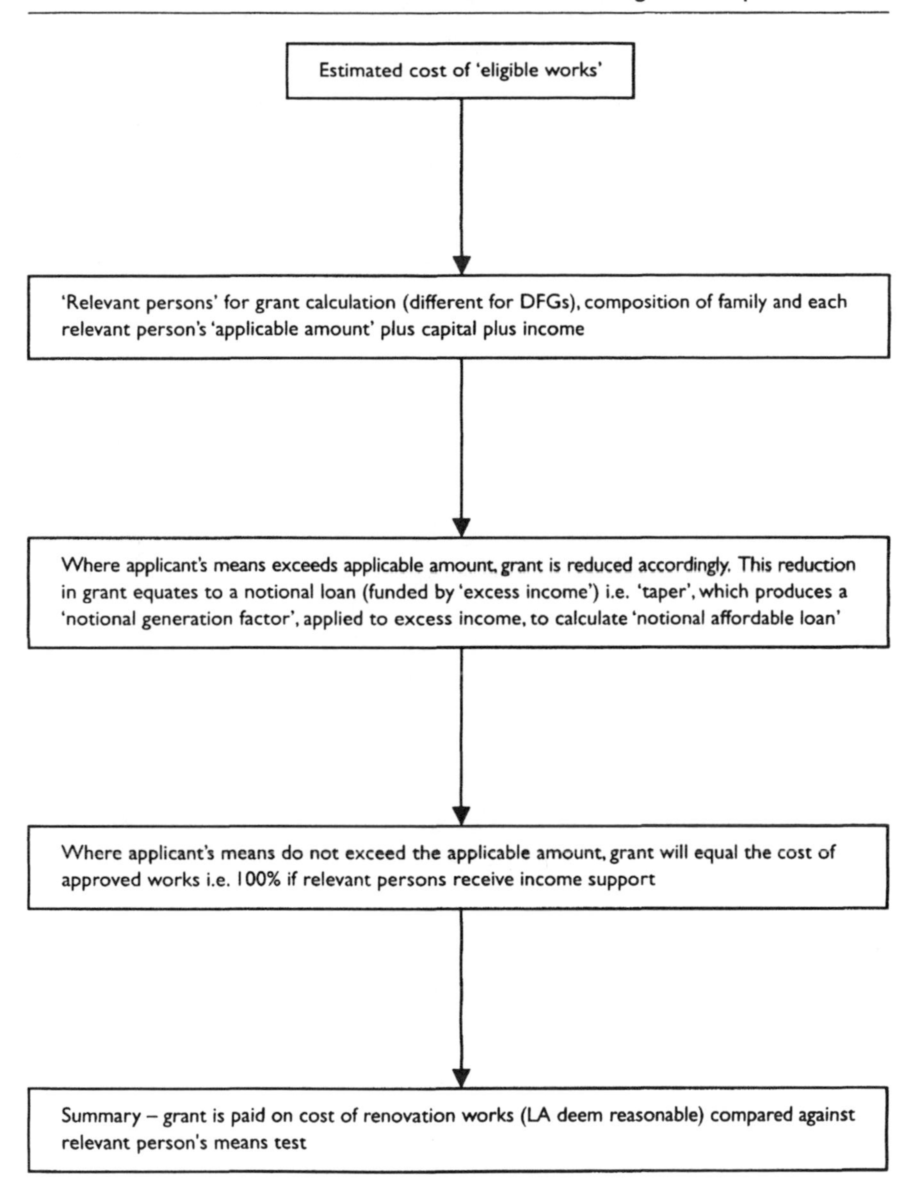

Figure 4.6 Calculation of a grant for owners and tenants

housing benefit, council tax benefit or disability working allowance. This does not apply for anyone aged 60 years or over. It is then up to local authorities to determine how to apportion their grant budgets.

Grant refusals and revision

There is no formal appeal against refusals, but local authorities should have systems in place to assist all applicants. This might include, for example, an independent officer determining the grant, or other possible courses of action if the house is unfit. These might include:

- provision of advice, e.g. funding, agency referral, list of builders etc.; and
- revision of a grant, such as where events are beyond the applicant's control, or in the case of unforeseen works.

The local authority can withhold payment of grant in some cases, such as when the original information provided was found to be inaccurate, or the works were not completed in 12 months.

A grant applicant may also make a complaint to the local government ombudsman if they were not satisfied with the grant administration. The ombudsman would investigate, and instigate appropriate action if the complaint were upheld because the local authority had failed to act properly. An ombudsman may investigate where a grant applicant were aggrieved being refused a grant outside of normal procedures, or if there were delay in processing the grant, etc. The investigation may find the local authority guilty of maladministration.

Auditing of grants

Control systems should be in place to:

- audit applications;
- follow grant process through;
- check contractors; and
- maintain quality of renovation works (although there is no duty to do so).

Grants and local housing strategies

The move to a discretionary system of grant, with associated enforcement powers in the 1996 Acts, reinforcing the need for effective housing strategies. The strategy should be published and should clearly set out criteria for eligibility. Increasingly private sector housing strategies are designed to make best use of stock within available resources. This is explored further in Chapter 5.

Home improvement agencies

Agency services can be provided by a variety of public, private or voluntary organisations to provide many forms of assistance to meet client need. There has been a growth in agency services in recent years, which started largely in the voluntary sector to promote better services, and more recently has received mainstream funding from local authorities in delivering their housing strategies. Many local authorities now offer clients an agency service to assist with their grant application. The nature of the agency, and the service it can offer, varies between authorities, based largely on available funding. Some have in-house services, whilst others contract agency services to a specialist housing association.

There are many reasons as to why agencies have gained momentum, not least through increasing government support and funding potential, particularly since the Local Government and Housing Act 1989. Agencies seek to cater for the specialist needs of an ageing population, a rise on owner-occupation and poor housing conditions which tend to be over-represented in this sector. Agencies tend to target particular vulnerable client groups, such as elderly or disabled people, with whom they have specific and specialist experience. They may offer a mixture of renovation, disabled facilities or home repair assistance grants, to meet the client's needs as well as assisting with the entire grant procedure from start to finish, which can be daunting to many people.

The Department of the Environment (DoE 1996), describes the many advantages of agencies run by local authorities and other organisations, such as Shelter's 'Care and Repair', and Anchor's 'Staying Put', as including the following:

- Support for sometimes disadvantaged applicants who are unfamiliar with administration, employing builders, building standards, and often where substantial finances are involved.
- Defining and enforcing quality standards.
- Providing the client and contractor with the security of knowing that everything will be dealt with properly.
- The capacity to plan work programmes and expenditure profiles.
- The promotion of community care policies by enabling people to remain in their homes.

Agencies have a valuable if not vital role in meeting a client's need through grant assistance and quality of service delivery in a way that traditional approaches can find difficult due to demands elsewhere. They can play a role in regulating standards through using known contractors, supervising the quality of the works, being supportive to clients and having a knowledge of other relevant organisations for referrals. Increasingly, in a climate of decreasing expenditure on grants, it is likely that their role will continue to assist in identifying alternative sources of finance, particularly private finance, which is set to play an increasing role in private sector housing renewal. An example of how a client's housing and social needs were met by an agency is illustrated in Box 4.2.

Box 4.2 Theory into practice – the role of home improvement agencies

Elsie, an owner-occupier, had been living alone since her husband had died some years previously and she had no immediate family who could help her. Her house had been built in the 1940s and had never had any major works carried out, but her late husband had always arranged routine repairs and decoration. She now had problems with lead pipework that had started to leak from the bathroom and into the kitchen below. She called in some builders who had pushed a leaflet through her door – they came and took a deposit and never came back to do the repairs, and they put the phone down every time she rang them.

Elsie was embarrassed to ask for help – she had never asked the government for anything and was anxious not to do so now, because she did not want to be a burden. However, she had no savings and only her pension, and the leak was getting worse. Eventually, she wrote to the council to see if it could help. The council put her through to the local care and repair agency. The agency dealt with all potential grant applicants aged over 60 years who may require specialist assistance and support.

The care and repair officer visited immediately on hearing that she lived alone and that her lead pipes were leaking, and because she was worried about the electrics. They sat with Elsie and filled in the relevant application forms, and read it back through to her before she signed it. They reassured her that the cost would be met in full through home repairs assistance, and arranged for a contractor to visit the next day to carry out emergency repairs to the worst section of the pipe. This began to reassure her, but she was still worried about the cost, not really believing that she was eligible for any financial assistance. Someone else visited from the agency and found that Elsie was eligible for income support, which she was not claiming, and sorted this out for her. They also suggested that she consider a home help, but she did not want to at the time.

The contractors visited at the arranged time, just after Christmas, as Elsie had requested and carried out the additional works the care and repair officer had identified as requiring attention. This included renewing the lead pipes, providing a smoke detector and chain to the front door and a new tank in the loft so that the maximum of £2,000 home repair grant was fully spent. She also had her loft insulated through the HEES scheme, which financed separately from her home repair assistance allocation, maximising her eligibility.

She was delighted with the works, her increased monthly income and having a bit of company to make a cup of tea and cake for.

Summary

- All grants are currently discretionary except for disabled facilities grants.
- There in an increasing need for local authorities to develop well-informed housing strategies.
- Grants legislation and supporting circulars change rapidly and reference should be made to these for up-to-date information.
- Agencies are becoming increasingly popular in delivering grants and associated services.

4.7 Disabled facilities grants

Introduction and legislation

The Local Government and Housing Act 1989 introduced the first grant specifically addressing a disabled person's needs amended by the Housing Grants, Construction and Regeneration Act 1996, disabled facilities grants are now the only mandatory grants, seeking to target further available resources. The Act is backed-up by guidance in DoE Circular 17/96 Annex 1 (DoE 1996). The demand for disabled facilities grants continues to rise, particularly because of community care policies, an ageing population, and growing awareness and expectations, which are set against a background of limited staff and funding resources trying to manage a complex system (Winders 1997).

A main aim of policy is that, wherever possible, people should stay in their own homes, with suitable care and support to live as independently as possible. Disabled facilities grants form a key component of this. Sometimes, however, a person with disabilities is no longer capable of living alone, or the accommodation may be unsuitable for various reasons, and, if so, rehousing may be more appropriate.

Definitions

The Housing Grants, Construction and Regeneration Act 1996 defines a person as disabled if they are or have:

- Sight, hearing or speech is substantially impaired.
- Mental disorder or impairment of any kind.
- Physically substantially disabled by illness, injury, impairment present since birth, or otherwise.

The definition is split between those above 18 years who the welfare authority considers to have disabled person's welfare needs under the

National Assistance Act 1948, and those below 18 years, where social services consider the child to be disabled under the Children Act 1989.

Application

Disabled facilities grants are available for a range of works to help a person with disabilities live more independently in their own home. There are two types: mandatory and discretionary. The main features of disabled facilities grants are summarised in Table 4.4.

Mandatory disabled facilities grants are available for essential adaptations and to enable use, or provision of, facilities in the home. DoE (1996) describes the type of eligible works as comprising the following:

- To make it easier to get into and out of the dwelling by, for example, widening doors and installing ramps.
- Ensuring the safety of the disabled person and other occupants by, for example, providing a specially adapted room in which it would

Table 4.4 Key criteria for disabled facilities grant eligibility

Type of grant	*Mandatory disabled facilities grants*	*Discretionary disabled facilities grants*
Purpose	For essential adaptations for movement into and around the home, and provision of essential facilities	For accommodation, welfare or employment needs of the disabled person
Residence qualification	None	None
Eligibility and restrictions	• Owner-occupiers, tenants and licensees • Local authority and housing association tenants	• Owner-occupiers, tenants and licensees • Local authority and housing association tenants
Maximum payable	£20,000 as a mandatory grant, discretionary above this total	Not applicable
Certificates and restrictions	At the discretion of the local authority	At the discretion of the local authority
Test of resources	• Person with disability and their spouse; or • Parents of a person with a disability, where under 18 years of age	• Person with disability and their spouse; or • Parents of person with disability, where under 18 years of age

Source: based on Housing Grants, Construction and Regeneration Act 1996 be safe to leave a disabled person unattended or improved lighting to ensure better visibility.

- To make access easier to the living room.
- By providing or improving access to the bedroom, and kitchen, toilet, washbasin and bath (and/or shower) facilities, e.g. by installing a stair lift or providing a downstairs bathroom.
- To improve a heating system to suit the needs of the disabled occupant.
- To adapt heating or lighting controls to make them easier to use.
- To improve access and movement around the home to enable the disabled person to care for another person who lives in the property, such as a spouse, child or other person for whom the disabled person cares.

Discretionary disabled facilities grants are available to meet accommodation, welfare or employment needs. DoE (1996) suggests that such works can include providing a safe play area for a disabled child or providing or adapting a room to provide a workplace for a disabled person. Local authorities also have the discretion to increase a grant above the mandatory level of £20,000. Such grants are available to occupiers, tenants or licensees with a relevant interest. Landlords, local authorities and housing associations can also apply on behalf of a disabled tenant. Local authorities should normally ensure that the house is statutorily fit, and a renovation grant may also be appropriate.

The local authority needs to ensure that the proposed works are necessary and appropriate, as well as reasonable and practicable. The local authority must, therefore, consult with welfare authority, normally an occupational therapist, about the adaptation needs and it is important that this service is well coordinated to help ensure a seamless service. The local authority needs to be satisfied that the proposed scheme is viable before approving a grant. There is no minimum age of the property and different grant conditions apply.

The means test is different from that of other grants because it centres on the disabled applicant. The local authority has some discretion in calculating a disabled facilities grant. The means test is based on the disabled person, and their partner's, income and capital. For a disabled child, it is based on the parents' means. Put simply, the test seeks to target grants, based on the difference the applicant is deemed able to afford, and the total cost of eligible work. The amount of grant is the amount needed to meet the costs of works above the applicant's contribution. This is the only mandatory grant, so local authorities can delay payment for up to 12 months to help manage their budget.

In some cases, particularly where works are small-scale, home repairs assistance may provide a quicker and more relevant form of assistance, which benefit the applicant with a lesser range of conditions than other housing grants. Such grants are consistent with wider care in the community and welfare moves under the Chronically Sick and Disabled Persons Act 1970 and is often administered through agencies, which provide appropriate support (Forrester 1998). In cases where applicants still struggle to meet their contribution, social services departments, and the Family Fund – administered by the Joseph Rowntree Foundation – can sometimes help (Winders 1997).

Disabled facilities grants – a streamlined approach

Circular 17/96 (DoE 1996a) emphasised the government's commitment to providing a strategic approach to dealing with disabled facilities grants at local level, and most housing authorities hold regular meetings with social services to develop strategies and maintain and improve close working relationships. A good strategy, with sound management, is needed to ensure a well-defined and consistent approach across the various organisations involved at all levels. This requires close working relationships between housing and social services departments, as well as agencies and voluntary organisations, which are involved locally. Whilst the theory is laudable, there are many practical issues to be addressed, including potential conflict between relative priorities, policies and in some cases departmental rivalry (Winders 1997). A summary of good practice in delivering disabled facilities grants is included in Table 4.5.

Joint working can be positive in addressing issues such as grant prioritisation, waiting times for initial visits and referrals, or where home repair assistance can be granted, and there is a need for discussion and agreement by all organisations involved so that the service can be effective and efficient as possible. There is also potential to keep a joint register of adapted properties. Unfortunately, despite the potential, there is little evidence to suggest that disabled facilities grants are positively promoted, and that there remain many inconsistencies in their administration (Age Concern and RADAR 2000). However, there are a growing number of websites advertising the availability of disabled facilities grants.

Close working with builders can also be key to providing a good service. Adaptations need to be reliable and provided sensitively by good builders. Local authorities can have an approved list of builders who are known to provide a good service at an economical cost. The builders form a key part of the disabled facilities grants role, and prompt payment will help keep them cooperative and available for emergency works as and when necessary.

The applicant's views are also paramount in deciding the best way

Table 4.5 Summary of good practice for disabled facilities grants

- Commitment to a strategy across all levels of all relevant organisations; committees that take a more strategic approach
- Creation of single team where everyone works together; establishment of mechanisms to resolve difficulties and promote good cooperation; well-documented policies and procedures for processing applications; joint training for social service and housing staff; joint visits for complex adaptations; provide applicant with an agreed and named contact point in relation to their grant works
- Regular liaison group meeting to discuss arising problems and possible solutions; constantly review the procedures for liaison
- Development of common data systems and of a register of adapted properties
- Provide more cost-effective adaptations; organise greater recycling of equipment
- Provide preliminary information on the grant contribution; give early information on eligibility, with a minimal wait for assessments and a swift, well-organised and supervised implementation after assessment provide loan advice to top up DFG
- Produce simple leaflets about the DFG system
- Product to be high quality and not ugly or embarrassing to enhance the quality of life
- Update information regularly on the budget and waiting times
- Consider using agency services
- Have mechanisms for consulting service users to learn about weaknesses and to act on suggestions; hold forums for people with disabilities and take their view into account; seek feedback on the service from groups as well as individuals; respect the values and priorities of the disabled person
- Follow-up those who drop out of the process, so all receive at least a minimum service, which does not exclude by means or prioritisation.

Source: based on DoE (1996) and Winders (1997)

forward. This includes determining the works required, as well as a proposed time-scale. Issues frequently require careful tackling in a cost-centred environment; whilst front line officers may like to maximise the grant-aided works available, this has to be balanced against budget constraints. There is also the issue of unnecessary 'labelling' of people with disabilities as 'vulnerable' through some very visible, cumbersome adaptations, particularly externally. Some disabled facilities grant applications are particularly sensitive, such as when someone is returning home for the first time from hospital after a disability caused by an accident, or where the applicant is terminally ill. Unfortunately, research suggests that many local authorities fail to consult with the applicant (Age Concern and RADAR 2000).

Summary

- Disabled facilities grants are on the increase.
- They are available on a mandatory basis in respect of access, and a discretionary basis in respect of accommodation, welfare or employment.

- There needs to be well-managed, cross-organisational strategies in place to ensure sensitive and appropriate grant administration by well-trained personnel.

4.8 Houses in multiple occupation – general introduction

Introduction and legislation

Houses in multiple occupation (HMOs) comprise some of the worst living accommodation, occupied by some of the most vulnerable members of society. Despite the poor conditions, local authorities were criticised by the Audit Commission (1991) for failure to develop strategies to tackle poor conditions. Numbers of HMOs are growing as local authorities struggle to maintain controls over existing, as well as newly identified HMOs.

Specific legislation governs HMOs because of their high-risk nature. Most is found in the Housing Act 1985 (as amended) Part XI and Regulations made there under. HMO legislation is currently under review, with proposals to introduce a mandatory licensing scheme.

This section examines dwellings that can be defined as an HMO, before looking at some of the unique issues surrounding this class of accommodation. Section 5.3 further examines how local authorities might develop strategies to deal with HMOs. It ends with a review of current proposals to introduce a mandatory HMO licensing system to add impetus to enforcing conditions in this sector.

Definitions

The first step in dealing with HMOs is to have a detailed understanding of what comprises an HMO. Precedent has frequently challenged the definition, so it is useful to divide the term into sections to help understand its scope.

The Housing Act 1985 (as amended) section 345 defines a house in multiple occupation as 'a house which is occupied by persons who do not form a single household'.

The definition comprises three parts:

- House.
- Occupied by.
- Persons not forming a single household.

House

'House' has no precise meaning, and so is a question of fact and law in the context of the Act and is further discussed in Section 3.5. The definition includes flats in multiple occupation. The Local Government and Housing Act 1989 extended the definition to buildings that may not otherwise be considered to be houses. For grant purposes, the definition excludes any part occupied by persons who form a single household. Relevant case law includes:

- Okereke v London Borough of Brent (1957) 1 Q.B. 42, where it was held that a house converted into self-contained flats could be considered as an HMO.
- Regina v London Borough of Camden, ex p. Rowton (Camden Town) Ltd (1983) 17 H.L.R. 28, QBD, where a purpose-built hostel was determined as an HMO.

Occupied by

Case law illustrates that this term can be applied flexibly. There does not have to be a particular legal arrangement such as a formal tenancy. Relevant case law includes:

- Minford Properties v Hammersmith (1978) 247 L.G.R. 429 D.C., where a tenant in receipt of a possession order was considered to be in occupation.
- Silbers v Southwark London Borough Council (1977) 122 S.J. 128, which determined that occupied broadly meant lived in.
- Thrasyvoulou v London Borough of Hackney (1986) 18 H.L.R. 370, CA, which determined that homeless persons placed by a homeless person's unit would fall within the definition.

Persons not forming single household

This element of the definition is a question of fact and degree. Relevant case law includes:

- Simmons v Pizzey (1979) A.C. 37, HL.
- Sheffield City Council v Barnes (1995) 27 H.L.R. 719, CA.

Simmons v Pizzey is a key case. In this instance, seventy-five women were living, temporarily, in a refuge. There was inconclusive evidence either way, so the case was determined on the basis of fact and degree, determining the premises as an HMO. The judge stated that three key factors should be taken into consideration in considering whether the occupants formed a single household, as follows:

- Size of the group and the number of persons.
- Stability of the group and frequency of changes.
- Relationship between members – had the group come together by accident or decision?

However, Sheffield v Barnes reviewed the basis of a shared house and its status as an HMO. In this case, five female students occupied a house, a Category B type under the CIEH definition. The judge considered Simmons v Pizzey, but added that each case was different and should be decided on its merits, suggesting that the following criteria were relevant:

- Origin of tenancy.
- Sharing of facilities.
- Occupation.
- Locks.
- Responsibility for filling vacancies.
- Size.
- Stability of group.
- Mode of living.

This case has caused surprise and some confusion within the environmental health profession (where HMO enforcement lies), and is still the subject of some contention. Local authorities need to be consistent in the way they assess and categorise HMOs. An example of some criteria that are used in identifying HMOs is summarised in Figure 4.7 – such criteria can be incorporated into an HMO survey sheet.

The definition can be further understood and applied in respect of national categories. The main recognised categories are listed by the Department of the Environment, Transport and the Regions (DETR) and by the Chartered Institute of Environmental Health (CIEH). The DETR uses these categories for national statistics, such as in the English House Condition Survey, or in research such as the ENTEC report (DETR and ENTEC 1997). They are:

- Bedsits.
- Shared houses.
- Households with lodgers.
- Purpose built HMOs.
- Hostels and houses used as care homes.
- Houses converted into self-contained flats.

CIEH categories (CIEH 1994) are broadly similar, but in general are more concerned with living arrangements than the DETRs. These are:

A. Contractual Arrangements	Y	N	Comments
Does each occupier have a separate contract with the landlord?			
If someone wants to move in or out do they have to deal only with the landlord?			
If someone moves out, do the remaining occupants have to find a new occupant?			
If someone moves out, do the remaining occupants have to pay the remaining rent?			
Who advertises vacancies?			
Is the property run by the Head Tenant as a business and for profit?			
Are the occupants granted exclusive possession of: (a) their own rooms or parts of the house? (b) are they lockable?			
Which parts are shared?			
Are household bills shared?			
B. Domestic Arrangements			Where? Evidence?
Are there separate washing and sanitary facilities in occupants' rooms or on separate floors?			
Do occupants make separate arrangements for cooking and eating?			
Are there separate cooking facilities in each occupants' room or on separate floors?			
Do occupants eat together when convenient?			
Is payment made on a board and lodging basis?			
How is the cleaning done? Do the occupants clean their own rooms and the landlord clean the common shared areas?			
Are there locks on bedroom doors? Are they used? Were any doors locked at the time of the visit?			
Do occupants have their own separate personal cooking utensils or are they shared?			
C. Permanency of Residents			Any evidence?
Do occupants come and go at frequent intervals or are occupancies on a long and stable basis?	F	S	

Is there any other evidence that the house is or is not multiple occupation?

Are the tenants prepared to appear in Court to give evidence regarding if necessary?	Yes?	No?

Source: based on CIEH (1994)

Figure 4.7 HMO checklist

- Category A – occupied as individual rooms, bedsits, etc., with some exclusive use and some sharing of amenities.
- Category B – houses occupied on a shared basis.
- Category C – some degree of shared facilities, occupation allied to employment.
- Category D – hostels, guest houses, bed and breakfast establishments.
- Category E – registered care homes and similar establishments (vulnerable people).
- Category F – buildings converted into self-contained flats.

Application

As can be seen from the range of case law above, the definition of an HMO can be elusive and this is perhaps the main reason that there are such differences in application of legislation. An example of assessing a property as an HMO is illustrated in Box 4.8. Clearly until a dwelling is defined as an HMO, no HMO-specific legislation can be applied and enforced. Most of the literature suggests a difference in statistics on HMOs, and there certainly remain concerns about consistency of enforcement (Audit Commission 1991). An example of the issues encountered in one case in initially defining a premises as an HMO is illustrated in Box 4.3.

Scale and nature of the problem

The (then) Department of the Environment (DoE 1996) calculated that some 640,000 buildings could be considered HMOs, accommodating over 3 million people. HMOs had increased in number since 1985. Nearly 60 per cent were in South East England, mostly in Greater London, with more than 80 per cent being in the private sector. Despite poor conditions, only one-quarter of local authorities had registration schemes. There are variations in statistics according to the source.

A major issue is being able to define a premises as an HMO in the first place. Whilst student accommodation was traditionally categorised as such, the precedent of Sheffield v Barnes 1995 threw this into confusion. This has become increasingly important as student accommodation has grown in recent years, and it is an important issue in university areas. There have been some high-profile cases of students being injured or killed in such accommodation, e.g. from carbon monoxide poisoning, in recent years, which has served to highlight some of the dangers in such accommodation, not just to students. Some local authority environmental health departments work closely with university accommodation officers to try informally to secure improvements where enforcement action would not be possible.

Box 4.3 **Theory into practice – is the premises a house in multiple occupation?**

The question of what can define a house in multiple occupation can sometimes be elusive, but sometimes an initial feel about a place, backed up by later firm evidence such as whether the residents are, or are not, living as a single household, can often turn out to be correct.

One particular house looked suspiciously HMO-like from the outside. It was detached, set back from the road and always seemed to have different cars parked on the concrete at the front, but it was not a hotel. There had never been any contact with the local authority, so there was no case history to scrutinise. Of course, the 'feeling' about the place that suggested that it was an HMO was not enough in itself to instigate HMO enforcement action – more evidence was needed, so a visit was arranged with the landlord.

Inside, there was an extremely strange, creepy feeling and several men there, who said they all lived in the building. This feeling was aggravated by the fact that all the tenants present seemed to be young men, but their relationship with one another was not clear, despite the EHO asking questions in attempt to assess if this could be defined as an HMO. Every question seemed to be met with a blank stare, and the landlord, who lived in the self-contained basement, remained shifty, but was willing to provide a bit more information. He said that he always housed young men who were ex-convicts, because no one else would take them in, so he had a lot of people coming and going and did not get to know them all very well, but some had stayed for years, and others sometimes got their mates to move in. He said that they made their own arrangements for meals and sometimes kept their doors locked, but did not understand why, because all their things would be quite safe – he always kept an eye out for the lads.

The EHO took careful notes of the verbal and anecdotal evidence obtained on the standard survey sheet, concluding that the property was in fact in multiple occupation and proceeded to prepare and serve relevant Housing Act 1985 (as amended) notices to bring it to required standards.

HMO works aside, there seemed to be another agenda going on. The landlord was a fraction too eager to allow officers access and arrange for what amounted to extensive and expensive works to be carried out extremely quickly and thoroughly. The builders confided in the EHO that they also had the creeps. They had found boxes of leather face-masks and whips in the loft, and were keen to get the works done quickly so that they could leave. On a revisit to assess progress with the works, there was a large projector screen and video set up in the basement, so, feeling vulnerable, the EHO did not stay down there for too long.

Once the works were finished, to an impressively high standard, everyone was pleased to depart until the interim risk assessment was due.

There is general agreement that the worst housing conditions are found in HMOs. A number of pressure groups, traditionally opposed to such accommodation, have now organised themselves in an attempt to improve such accommodation for some of the most vulnerable members of society where, frequently, alternate accommodation is not available. Many local authorities, criticised by the Audit Commission (1991) for lack of HMO enforcement action and strategies, have now taken action to promote conditions in the poorest sector of housing, but there is still a long way to go. This is explored further in Section 5.3.

HMOs – proposals for licensing

It is often said that a licence is needed to run a kennel, but that any idiot can run an HMO. In many senses this is true, but the reality of dealing with residential accommodation in the market place in a climate of increasing homelessness makes the situation more complex. HMO licensing is being considered at the time of writing.

Conditions in HMOs are commonly found to be amongst the worst in housing, with increasing numbers of low-income groups occupying them (DETR 1999). There is an inherent conflict between poor conditions in HMOs and their important role in providing low-cost accommodation. It is this issue that is key to the debate surrounding proposals for HMO licensing. The government is keen to regulate HMOs, yet not to be over-burdensome as to prompt their withdrawal from the market place. This has led to some delay in the scheme's implementation.

Discretionary HMO registration schemes have not proven popular with local authorities, and the numbers of schemes established have remained low. The proposal to introduce a mandatory licensing scheme seeks to increase the impetus to deal more proactively with HMOs. A mandatory rather than the current mainly discretionary HMO legislation would require that local authorities prioritise their resources to this area. Initially, there is no doubt that such a scheme would involve a considerable resource input, and additional staffing may be required in some areas. There are many issues to be addressed, such as which HMOs would be included, how the scheme should be administered, what the responsibilities of the licensee might be, how the scheme would differ in administration from the current position, and so on. If not properly thought through, there is a risk that it could become overly bureaucratic and have little overall impact on HMO conditions.

The general proposal for HMO licensing has been favourably received by local authorities and the CIEH (1999) as a long-term way forward in controlling conditions in HMOs. It is seen to enable a more accurate profile of HMOs locally and nationally, and to add credibility to control powers through a proactive, streamlined, risk-assessed approach.

Discussion remains on-going on the best way forward, and it is too early to comment in detail on the likely legislation.

Summary

- The definition can sometimes need breaking down into component pieces to apply so that HMO specific legislation can be enforced.
- HMOs comprise the worst living conditions, frequently occupied by the most vulnerable members of society.
- HMOs can be difficult to locate, so statistics remain in question.
- Although local authorities are increasingly developing strategies, there is still a long way to go in addressing a growing number of HMOs.
- The government is currently proposing to introduce an HMO-licensing scheme to improve conditions across the worst housing sector.

4.9 Houses in multiple occupation – amenities and occupation control

Introduction and legislation

One of the main issues in HMOs is the number of amenities available to tenants. Amenities are concerned with the number of personal washing, toilet and kitchen facilities available to tenants, and their application is discretionary. The CIEH (1994) provided guidance on the number and location of amenities expected based on HMO category. Some local authorities have adopted these directly, whilst others have amended them or developed their own. Overcrowding in HMOs is included here to illustrate the difference between overcrowding and over-occupation.

Relevant legislation is the Housing Act 1985 (as amended) section 352, which is concerned with 'fitness for the number of occupants'. This standard is in addition to the statutory standard of fitness (under section 604). HMOs are required to have:

- satisfactory facilities for the storage, preparation and cooking of food, including sinks with hot and cold water;
- adequate numbers of suitably located toilets;
- adequate numbers of suitably located baths/showers and wash hand basins with hot and cold water;
- adequate means of escape in case of fire; and
- adequate other fire precautions.

This section just looks at amenities. Means of escape in case of fire and other fire precautions are considered in Section 4.11.

Application

The terms 'suitable' and 'adequate' are based on DoE Circular 12/92 (DoE 1992), and a local authority's own determination of numbers for fitness. CIEH (1994) generally suggests one set of amenities per five occupants or households. It emphasises the need for flexibility based on local housing conditions.

There is a discretionary power under section 352 where HMOs are considered 'not reasonably suitable for occupation'. The decision to enforce is based upon inconvenience, nature of defects or deficiencies, probability of accident, damage to health or fatalities and the severity, extent or location of those defects. DoE 12/92 (DoE 1992) and CIEH (1994) refer to the general principle that different standards of provision should apply to different categories of HMO. In addition, local authorities are encouraged to adopt their own standards, but ones based upon this guidance.

The basis of amenity standards is concerned with hazards and risks associated with multiple occupied dwellings. Local authorities should have regard to the:

- type of provision and location of facilities;
- scale of provision;
- suitability of sinks and hot and cold water supplies, cookers, worktops and food storage facilities;
- layout of kitchen facilities; and
- lighting and ventilation of kitchens.

It is desirable for each household to have facilities on the same level as the unit of accommodation, but this is not always possible. There should be adequate numbers for tenants to cook at reasonable times. Where facilities are shared, there should normally be no more than three households per facility, or five where the HMO comprises more of a shared household type arrangement. Partitioned food storage space should be provided. Facilities should be reasonable and be capable of being cleansed.

Toilets and personal washing facilities should be adequate for the convenience of occupants in terms of privacy and personal hygiene. Decisions should be based on the nature and layout of the HMO, ideally within the unit of accommodation, but realistically this is not always possible. Facilities should be accessible without going through someone else's accommodation. DoE 12/92 (DoE 1992) requires that the toilet be separate from the bathroom where there are more than five occupants. A wash hand basin should be provided in each unit of accommodation. Cubicles should have lighting and ventilation, provided by natural or artificial means. Facilities should be capable of being effectively cleansed, with slip-resistant floors.

If an HMO lacks adequate amenities based on the number of occupants, it is considered to be 'over-occupied'. This is distinct from the overcrowding definition, which is based around the size of occupation. Over-occupation is related to the number of facilities available.

For a flow chart on the procedure, see Bassett (1998: 112) and Figure 4.8.

Control of over-occupation – Direction Orders

Control of over-occupation is by a 'Direction' under the Housing Act 1985 (as amended) section 354. The purpose of Direction is to secure, by Order, the number of individuals and households in an HMO. Directions can only be used in relation to section 352 matters. They cannot be used to limit occupation of a particular room by space, but a zero Direction can be applied where facilities are lacking. They cannot be used to revert house to single occupancy.

The general principles of Directions are:

- Decide whether to consider the HMO premises in whole or in part.
- Consider the number of toilets and bathrooms and their location – for determination of the number of individuals.
- Consider the number of kitchen facilities – for determination of number of households.
- Consider room areas and space standards – should an overcrowding notice also or alternatively be served?

A Direction prohibits occupation above the number directed. Reduction is normally achieved by 'natural wastage', so that no one is made homeless.

It is an offence to fail to comply with a Direction, and service is a local land charge. Following service, a local authority can require information occupation by number, the number of households, names, rooms used and so on. It is an offence not to provide information. Adherence to Directions requires regular monitoring that requires dates, names and so on of new occupiers, but tenants may be wary of giving information for fear of reprisal.

Directions can be used to prevent situations arising or remedy existing situations. It is necessary to consider effect of availability of low-cost accommodation in the area if its use is too heavily restricted by local decisions and strategies. The numbers of prosecutions and fines are low.

For a flow chart on the procedure, see Bassett (1998: FC116) and Figure 4.8.

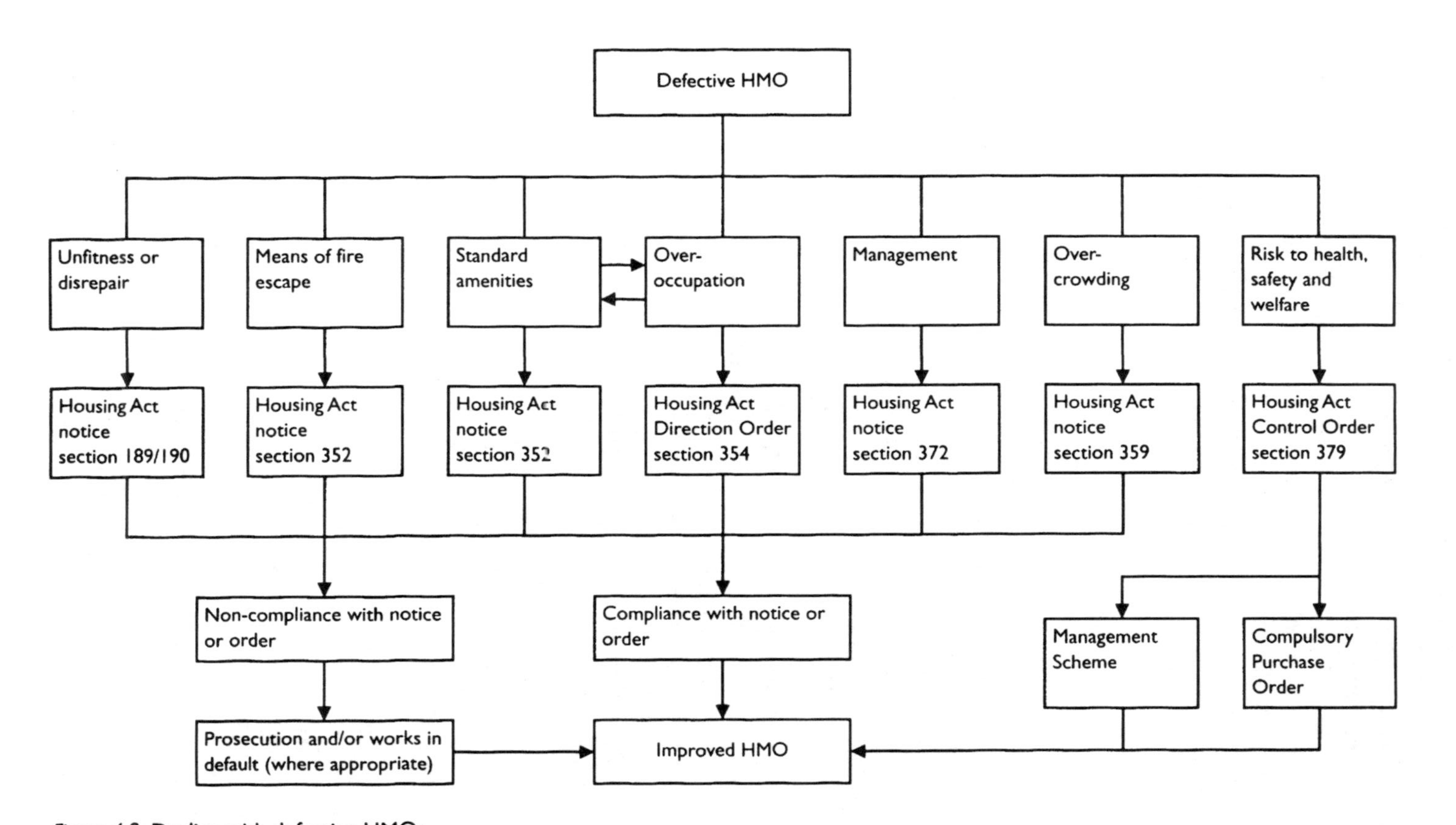

Figure 4.8 Dealing with defective HMOs

Control of overcrowding in HMOs – overcrowding notices

Remedy for overcrowding in HMOs is found under the Housing Act 1985 (as amended) section 358. Overcrowding in HMOs can be controlled on a room-by-room basis, e.g. in hostels, hotels and so on. Local authorities have discretionary powers to serve notice where:

- there is an excessive number of people on the premises, having regard to the rooms available; or
- it is likely that an excessive number will be accommodated.

The person served is to prohibit occupation of sleeping room in excess of permitted number (including new residents), and rooms are to be occupied in such numbers so that it is possible to avoid sexual overcrowding (including new residents). They are also required to provide on request details of occupation, such as the number of individuals, households, names, rooms used and so on. The person served can appeal against overcrowding notice, which may be revoked or varied.

For a flow chart on the procedure, see Bassett (1998: FC113) and Figure 4.8.

Summary

- Local authorities have discretion to require amenities in HMOs, with levels of provision normally based on HMO category, based on a ratio of occupants and/or households to amenity.
- Control of over-occupation (i.e. numbers per amenity) is by way of a Direction.
- Over-occupation is distinct from overcrowding.

4.10 Houses in multiple occupation – standards of management, control and registration

Introduction and legislation

This section looks at some key legislation governing the management, control registration of HMOs. This is found under the Housing Act 1985 (as amended) and regulations made thereunder and is referred to in the body of the text.

Definitions and application

Management, control, registration and miscellaneous legislation are considered in turn.

Management Regulations

The Housing (Management of Houses in Multiple Occupation) 1990 placed responsibilities on the person managing the HMO to ensure proper standards of management. The regulations apply to all HMOs and failure to comply with them is an offence, but the local authority first needs to show that the manager was aware of responsibilities under the Management Regulations.

The regulations are summarised in Table 4.6, but in general require the HMO manager:

- To ensure repair, maintenance, cleansing and good order of water supply and drainage; common parts and facilities; living accommodation; windows and ventilation; means of escape in case of fire and other fire precautions; outbuildings, yards, etc.
- To make satisfactory arrangements for refuse disposal; take reasonable precautions to protect general safety; display a management notice; to provide information regarding occupancy.
- There are also duties on the occupants to enable the manager in the above.

Application is by service of a Housing Act 1985 (as amended) section 372 notice, for neglect of management. The local authority can prosecute for contraventions directly, or serve notice requiring necessary works. The 'minded to' notice procedure applies. The notice should be served on the manager, with a copy to persons with an interest. The local authority can prosecute for failure to comply and/or instigate works in default.

Table 4.6 Summary of HMO Management Regulations

Manager's duties	Regulations require managers to take a wide range of action to protect the safety and welfare of the residents of the house
Water supply and drainage	The water supply (including tanks and pipes) and drainage system must be kept clean and be in good order. Pipes, cisterns and other fittings must be protected from frost. There must be no unreasonable interruption to the water supply
Gas and electricity	There must be no unreasonable interruptions of the gas and electricity supplies
Parts of the house in common use	All common parts of the house must be kept clean, be in good order and be properly decorated. Stairs, passages, lobbies, balconies, etc. must be kept free from obstruction and be in good repair. Floor coverings must be safe and properly fixed
Installations in common use	Installations for making use of gas and electricity in common use such as lighting and heating, shared fittings such as toilets, baths, wash hand basins, sinks must be properly maintained and kept clean.

	The lighting to common stairs, landings, passages, shared bathrooms, toilets, etc. must be adequate. Letterboxes and other installations for receiving mail have to be properly maintained
Living accommodation	At the beginning of a resident's occupation, managers must make sure that the accommodation is clean. Thereafter, resident's accommodation, including gas, water, electricity, lighting, heating and sanitation must be kept in good repair
Windows and ventilation	All windows (whether in flats or common parts) must be maintained in good repair and be in working order
Means of escape from fire	All means of escape (e.g. fire doors and external escapes) and all associated equipment (e.g. extinguishers, alarms and emergency lights) must be maintained in good order and repair and be kept free from obstructions. Readily visible signs indicating all means of escape must be displayed in the house. NB – alarms and warning systems must be tested at the officially recommended intervals and a written record of testing should be kept
Outbuildings, gardens, etc. in common use	All common outbuildings, yards, areas and forecourts must be maintained in repair, a clean condition and be in good order. Gardens in common use must be kept in reasonable condition. All boundary walls, fences and railings must be kept in a safe condition
Disposal of refuse and litter	Refuse and litter must not be allowed to accumulate unreasonably. Suitable and sufficient bins or other receptacles must be provided for storing refuse and adequate arrangements must be made for the disposing of refuse
General safety of residents	The manager has a wide general duty, having regard to the design and structural condition of the house, to protect residents from injury as a result of those conditions
Display of notices	A notice stating the name, address and telephone number (if any) of the manager(s) of the house must be displayed where it can readily be seen by residents. The manager must take reasonable steps to ensure that the notice remains on display and is updated as necessary
Duty to give information to the local authority	If required in writing to do so, the manager must give the local authority details about: • the number of individuals and households in the house; • the number of individuals in each household; and • the purpose for which each room in the house is being used
Duties of residents	Residents have duties to: • avoid hindering or frustrating the manager in performing his or her duties; • allow the manager, at all reasonable times, to enter the (resident's) accommodation when it is necessary for performing the manager's duties (e.g. for carrying out repairs); • provide reasonable information to the manager; • comply with reasonable arrangements for means of escape from fire, fire precautions and for storing and disposing of refuse and litter; and • avoid damaging anything the manager is obliged to keep in repair • NB – landlords are not required to make good, repair or maintain anything that a tenant is entitled to remove from the house

Souce: based on Housing (Management of Houses in Multiple Occupation) Regulations 1990

For a flow chart on the procedure, see Bassett (1998: FC114) and Figure 4.8.

Control Orders

Control Orders provide for local authorities to tackle conditions in the worst of the worst of HMOs under the provisions of the Housing Act 1985 (as amended) sections 379–94. Control Orders enable a local authority to take control of an HMO for 5 years and bring it to required standard. These provisions apply to protect the health, safety and welfare of tenants and where conditions warrant action under legislation. This is a very rarely used provision, probably because of the major resource implications to local authorities.

Following a Control Order, a local authority can make a Compulsory Purchase Order (within 28 days) or prepare a Management Scheme within 8 weeks to manage the HMO for 5 years. This is a rarely used provision.

For a flow chart on the procedure, see Bassett (1998: FC117) and Figure 4.8.

Registration

The Housing Act 1985 sections 346–51 (as amended by the Housing Act 1996) provides the discretionary basis for local authorities to establish Registration Schemes in their areas. Their objective is to help identify and control conditions in this sector, but very few have been established nationally.

Local authorities can set up Registration Schemes for the whole, or part of, their area. Model Schemes are available from the DETR, and further information is provided in DoE 3/97. There are three main types of scheme:

- Notification scheme.
- Control scheme – where conditions can be attached.
- Special Control Provisions – where adverse impact on neighbourhood, or problems are anticipated as a result of HMO development.

Once registered, the HMO details are available on a public register. A registration fee can be charged.

For a flow chart on the procedure, see Bassett (1998: FC111).

Miscellaneous legislation

There is other legislation available to deal with poor living conditions in HMOs. This includes:

- Housing Act 1985 (as amended) – including repair provisions, grants, etc.
- Environmental Protection Act 1990 – statutory nuisance and Building Act 1984 section 76, accelerated procedure.
- Health and Safety at Work etc Act 1974 – general duties of care.
- Furniture and Fittings (Fire Safety) Amendment Regulations 1993 – enforced by trading standards officers.
- Gas Safety (Installation and Use) Regulations 1998 – enforced by the Health and Safety Executive (see Leaflets section page 278).
- Local Government (Miscellaneous Provisions) Act 1976 – reconnection of gas, water and electric supplies.
- Anticipated Duty of Care under the Housing Act 1996 section 73 – may now be postponed following a decision on HMO licensing.

Summary

- The Management Regulations provide a wide basis for tackling a range of poor conditions in HMOs. They can only be used to manage what is already there, not to require provision of something that is not.
- Control Orders provide the remedy to deal with the worst HMOs, but are very rarely issued, due to resource requirements.
- Registration Schemes can be set up to cover the whole or part of a local authority area, for which a fee can be charged. Model Schemes are available to assist local authorities.
- Although there is a mass of HMO legislation, other related legislation may present a more suitable remedy in certain cases.

4.11 Houses in multiple occupation – means of escape in case of fire and other fire precautions

Introduction and legislation

Provision of means of escape in case of fire and other fire precautions is a priority in HMOs. Ninety-eight per cent of local authorities consider means of escape as one of the main problems in HMOs and the statistics, though varied, point to a higher risk of fire and associated injury than other dwelling types. Estimates of risk of fire death vary from at least six times the likelihood than in other dwellings (DETR and ENTEC UK 1997), to up to ten times (Audit Commission 1991). The National Consumer Council (1990) has estimated that as many as 80 per cent of HMOs have inadequate means of escape in case of fire.

The increased risk of fire occurs because there are generally more kitchens, more electrical goods with heavy ampage equipment, poorly maintained electrical systems, mobile heating appliances, small rooms with clothes too near heaters, smoking, and so on. A frequent change in tenants combined with a change in electrical equipment, incorrect use and overloading of appliances, combined with DIY, adds to the fire risk.

Behaviour in fires is also paramount. Research suggests that occupants in lower storeys fail to warn those in upper storeys. Fire can spread rapidly through staircase enclosures if fire doors are not shut, or if fire protection is inadequate. The development of fires is extremely rapid. Temperatures of 180°C causes ignition, and slow smoking fires can be detected by automatic fire detection (AFD) at this early stage. Flashover occurs at 800°C, because the high temperature means other items in a room may ignite without direct contact with a fire. Evacuation must be early, pointing to the need to AFD, preferably with detectors in rooms where a fire is likely to start, if legislation permits this.

Key legislation for fire safety in HMOs is:

- Housing Act 1985 (as amended) section 352.
- Section 352 (iv) – provision of adequate means of escape in case of fire.
- Section 352 (v) – provision of adequate other fire precautions.
- DoE Circular 12/92 and BS 5839.
- Fire Precautions Act 1971 (enforced by the fire authority).

There are many bodies involved in domestic fire safety, including environmental health officers, fire safety officers, building control officers, social services and so on. One problem is that initial enforcement and on-going management can be complex, particularly in the residential sphere. The enforcement authority is not always clear-cut. Different legislation is not always in harmony, and not contemporaneous. In some HMOs, the fire authority may have responsibility for fire safety, whilst the local authority may have responsibility for all other housing legislation. There were proposals for the fire authority to become the main enforcer in HMOs, although it currently seems as though this will remain with local authority environmental health departments as this is likely to be more streamlined since they already enforce other HMO legislation. Good working relationships and communication are therefore crucial.

The risk of fire in bedsit accommodation is ten times higher than other designs. Detection should be widely installed, as:

- Forty per cent of fires start in kitchens.
- Sixty per cent of fire deaths occur in the room where the fire starts.
- Fifty per cent of fatal fires start in the living or dining room.
- Thirty per cent of fatal fires start in bedrooms, when the occupant is asleep.

It is important to have these issues in mind when deciding the type of AFD required, balancing a thorough system with the likelihood that the alarm might be ignored if it sounds too regularly in error.

Definitions

The Housing Act 1985 (as amended) section 352 provides the legal remedy for means of escape in case of fire and other fire precautions. 'Means of escape' refers to the structural aspects of a fire escape route, whilst 'other fire precautions' refers to matters such as fire fighting equipment.

Application

The general principles of 'means of escape' are to provide a protected route with a 30-minute fire integrity to allow safe escape in case of fire. For this purpose, horizontal and vertical fire separation is required. In some cases this is backed up by AFD, emergency lighting and fire fighting equipment (FFE). It is also important to ensure that travel distances are not exceeded.

Requirements for means of escape are laid down in Department of the Environment Circular 12/92 (DoE 1992). This circular requires that the following issues are considered:

- Adequacy of structural protection between occupancies and the protected route.
- Travel distances and layout of property.
- Nature of escape route.
- Need for fire precautions and fire fighting equipment.
- Confidence in management.

In terms of the practicalities of means of escape, there are some key points to consider. Generally speaking, a 30-minute fire protection is required horizontally and vertically between units and onto the protected route. The protected route is the way out to a place of safety. In some cases 60 minutes of fire protection is required, for example if the HMO is above a high-risk commercial premises. The manner in which structures are upgraded varies, but regard should be given to advice in DoE 12/92.

This section does not seek to provide thorough details in respect of means of escape, but provides simply an overview. Further information can be obtained from the References and Further reading suggested below. However, put very briefly, the following points are useful to consider.

Doors

FD30S means a 30-minute fire door fitted with hot and cold smoke seals. Existing doors can be protected from the risk (i.e. room) side to 30 minutes' fire integrity. The frame, architrave and stop should be capable of 30 minutes' fire integrity. Doors should be hung on three fire standard hinges, with gaps of no more than 4 millimetres. Overhead self-closers should be provided.

Walls, ceilings and floor

There are many possibilities and decision to upgrade should be based on what is already there but generally for 30 minutes' fire integrity the following should be provided:

- Stud-bearing walls/ceilings – 12.5 millimetre plasterboard, with joints taped and filled.
- 18 millimetre tongue-and-groove plywood or chipboard.
- 60 millimetre insulation fibre between joists can protect the attic space.
- Fire-resisting glazing (e.g. Georgian wired) – set as per fire doors.
- There are many other possibilities.

Along the protected route

- The protected route should be kept clear, containing nothing that might burn, including wall finishes, furniture and so on.
- Cupboards (e.g. containing older gas meters and storage) should be fully upgraded still to be used, or removed, or emptied and locked shut.

Automatic fire detection (AFD)

Automatic fire detection has become increasingly prominent in recent years, taking over from structural lobbies and so on. Its purpose is to provide early warning in the event of fire to enable residents to escape to a place of safety. Generally speaking, AFD should be provided as follows:

- Domestic-type smoke detectors up to two storeys.
- L2 system (BS 5839) for more than two storeys – this comprises a control panel, sounders, detectors and manual call points.

The choice of AFD depends on the risk of the premises, BS 5839 allows some flexibility in L2 systems. L2 systems are designed to cover the protected route and higher risk rooms, sometimes even cupboards, where fires are most likely to start. They are interlinked with a mix of mains and battery power so that all sound simultaneously. Heat detectors are provided in kitchens, but smoke detectors are provided elsewhere.

The control panel should be located at the entrance of the HMO. Since AFD systems can be zoned, this illustrates the layout so that emergency crews are immediately able to locate the fire, which is particularly important in larger HMOs. Alarm sounders should be capable of producing 75 dB(A) at the bed head.

Emergency lighting

The emergency lighting system should be independent from the main supply and should provide sufficient illumination to enable persons to see their way clearly out of the premises. It should operate on any failure in the normal lighting system. Emergency lighting should be sited in accordance with BS 5266: Part 1, both to indicate the route of the escape and to illuminate hazards. Illumination should be uniform, being included at each intersection on corridors, at each exit door, at a change of direction, at staircases, near each fire alarm call point, near fire fighting equipment, to illuminate exit and safety signs, and so on.

There are two main types of systems:

- Non-maintained, where lamps are illuminated when the main supply fails, energy from battery and control gear, charges in rest.
- Maintained where lamps can be illuminated when the mains supply is healthy and has a separate connection to power.

If AFD and emergency lighting are required, it is useful to ask for Completion Certificates for Installation or Alteration and require regular inspection, as well as regular testing under the Management Regulations.

Other fire precautions

- A 9-litre water extinguisher is normally required at the head of each stairwell and the final exit door.
- Carbon dioxide or dry powder extinguishers and fire blanket(s) are required in each kitchen.

General requirements for means of escape in case of fire and other fire precautions are summarised in Table 4.7.

Table 4.7 Summary of requirements for means of escape in case of fire and other fire precautions in HMOs

Type of accommodation	*DoE Circular 12/92 requirements*
Two storeys	Stairway to be a protected route; single-point mains-operated smoke detectors; L2 system if a hostel
Three or four storeys	Stairway to be a protected route; L2 fire-detection system
Five or more storeys	Protected route with an L2 system, plus either: • second internal escape; • upward and downward means of escape via another property; and • lobbies, with occupancy limited to five storeys

Source: based on DoE (1992)

Travel distances

Travel distances vary in each HMO, but some general rules apply. Assessment is carried out in two stages:

- A – within the unit of accommodation.
- B – from the unit to the final exit.

Assessment based on escape possibilities, and

- A – maximum distance from any point within a room to the room exit should not exceed 9 metres.
- B – maximum distance from the accommodation to the stairway should not exceed 7.5 metres.

Different provisions apply for inner rooms and access rooms.

There are additional requirements where a HMO is more than five storeys.

Applying the legislation

The Housing Act 1985 (as amended) section 352 provides for fitness for number of occupants. This incorporates both amenities (outlined in the introduction to Section 4.9) as well as means of escape in case of fire and other fire precautions.

HMOs can be extremely time-consuming to deal with. It is first necessary to carry out a full survey and identify necessary works. It can sometimes take several visits on different dates and at different times to gain access to all rooms. Scale plans must be drawn up, together with a specification of work based on the information above and DoE (1992). It is normal practice to highlight the protected route in red on the plans and illustrate relevant features such as the general location of AFD, FFE, etc. There is a duty to consult with the fire authority, and each local authority will have its own arrangements in place for this. The proposed specification may be agreed or amended at this stage. A minded to notice follows. It is only then that relevant formal notice(s) can be issued – in most cases, a section 352 notice would be served.

It may be that some parts of the HMO are so designed and potentially unsafe that they require formal closure by Order. The Housing Act 1985 (as amended) section 368 provides for closure provisions due to unsatisfactory means of escape. An example might be closing an attic room that has an inadequate way out – an example of this is illustrated in Box 4.4.

For a flow chart on the procedure, see Bassett (1998: FC112, FC115).

Recent legislation (Housing Act 1985 (as amended) section 365) has extended the mandatory duty to inspect and enforce in three-storey HMOs under the Housing (Fire Safety in HMOs) Order 1997.

Other legislation

There is further legislation outside of the Housing Act 1985 (as amended) that can provide remedy in respect of fire safety. This includes:

- Fire Precautions Act 1971 – which is useful to close dangerous rooms quickly.
- Building Act 1984 section 72 – can be used where there is no means of escape. The section applies where a building exceeds two storeys in height and the floor of an upper storey is more than 20 feet above ground level on any side of the building that is let in flats or tenement dwellings; is used as an inn, hotel, boarding house, hospital, nursing home, boarding school, children's home or similar; or used as restaurant, shop, store or warehouse and has an upper floor selling accommodation for persons employed on the premises.
- Housing (Management of HMOs) Regulations 1990, particularly section 10 – means of escape in case of fire and other fire precautions; section 13 – general safety of residents; and section 16 – duties of residents. These regulations are useful to ensure the good order and repair of means of escape and apparatus, the design and structural conditions are such to prevent injury, as well as to ensure that residents enable the manager in these functions.

Fire risk and the ENTEC Report

There has been some concern as to the nature and extent of means of escape and other fire precautions works required, and, as a result, the

Box 4.4 Theory into practice – fire safety in HMOs

This example shows how potential risks of harm from inadequate means of escape can be resolved before anyone is unnecessarily injured, or even killed, in a fire.

An initial visit to a house in multiple occupation revealed that one room was in the loft, and access to it was only possible by a ladder. The room itself was clad with some hardboard, had no windows, and the only furniture was an unmade bed and a chest of drawers. Officers explained that this room was not suitable in terms of means of escape from fire and would have to be formally closed and no longer used as a residence. The landlord said that, despite the bed, it was not currently occupied and he would not use it again. It seemed that in this case the situation might be relatively easily resolved, but it clearly needed formalising.

The officer called the fire safety officer, who agreed verbally that the best option was immediately to close the loft room formally, and further means of escape in case of fire and other fire precautions would be agreed at a later stage when plans and a schedule of works were ready for consultation. The room was immediately closed under the Housing Act 1985 (as amended), a copy of the order being posted on the sealed loft hatch. A revisit was made to ensure that it was no longer being occupied, which it was not.

When the major means of escape works were carried out, the landlord took the opportunity to redesign this end of the HMO, and created a self-enclosed, spacious bedsit from where the loft access had previously taken up space.

government commissioned research to investigate fire risk as part of a package to improve HMO safety (DETR and ENTEC 1997). The research drew together evidence on factors contributing to HMO fire risk as a basis for further research and review. It had several aims, including to identify factors responsible for fires, and how these might be exacerbated in HMOs; HMO types and features with high risk; local authority and fire authority roles; scrutiny of existing legislation and proposals for the future; prioritising enforcement activity; and identifying risks between DETR HMO categories. In doing so, evidence of HMO fire deaths by HMO category and fire brigade reports were examined.

The report made four major recommendations that remain under review:

- Housing (Fire Safety in Houses in Multiple Occupation) Order 1997 – that the existing duties for local authorities to enforce fire safety be increased to incorporate occupancy, not just risk arising from structural features, with a new scheme introduced for lower-risk HMOs.
- Approved Code of Practice – the draft code to be reviewed to incorporate increased fire protection (with relevant guidance) in high-risk HMOs and be related to occupancy, management and other risk factors; provision of secondary fire doors, additional AFD, changes in layout and provision of sprinkler systems to restrict fire and smoke spread; reduction of requirements in lower risk premises; alternate safety measures to be considered where occupants have special needs; increased emphasis on preventing fires.
- Fire safety management and enforcement – enforcement alone is not enough and education is also necessary. Training is recommended to improve fire safety management to reduce the number of fires to secure self-compliance; ensure consistency between regulatory regimes; place greater emphasis on fire prevention.
- Definition of HMOs – that the current wide definition be retained, as it already takes occupancy into account. The risk assessment could determine level of fire precautions required.

Summary

- HMOs have a higher risk of fire than other residential accommodation.
- There is specific legislation governing fire safety in HMOs.
- The current standards are under review.

4.12 Insanitary premises and related issues

Introduction and legislation

Dealing with insanitary premises – more commonly referred to as 'filthy and verminous premises' under the Public Health Act 1936 – comprises one of the most basic public health issues facing EHOs. Premises become filthy and verminous for many different reasons, including self-neglect. There are no national statistics on increases in filthy and verminous cases, but cases are increasing. Whilst filthy and verminous conditions are not exclusively confined to those with mental health problems, there is no doubt that the rise is at least partially linked to care in the community policies, where care services can sometimes fail to operate effectively outside of psychiatric hospitals (Barham 1997).

Addressing filthy and verminous conditions can range from being challenging to outright distressing and disturbing, and often requires a multi-agency approach in an attempt to find a satisfactory resolution. EHOs are frequently frustrated by the fact that their public health enforcement powers are a 'service' of last resort which tends to be a short-term fix and likely to require review in a matter of months.

Remedy for such conditions is found in the Public Health Act 1936 as amended by the Public Health Act 1961 and used alone, provides a draconian response. Identification of persons living in filthy and verminous premises often means that there are other social needs to be addressed, although sometimes the clearance is all that can be reasonably achieved. The National Assistance Act 1948 first provided for client care needs and a multi-agency approach is increasingly encouraged to find solutions in such cases. This section also reviews the Public Health (Control of Disease) Act 1984, which contains most of the current provisions relating to infectious disease control.

Definitions

The Public Health Act 1936 does not define 'filthy or unwholesome'. It does, however, define 'verminous' in its application to insects and parasites, which includes their larvae and pupae. The definition of what comprises a filthy and verminous premises may appear subjective, although officers have little doubt when faced with such conditions – and the photographs in Boxes 4.5 and 4.6 make it clear.

Application

Where a local authority is satisfied that any premises

- are in such a filthy or unwholesome condition as to be prejudicial to health, or
- are verminous

they have a duty to give notice to the owner or occupier requiring remedy. This includes, where appropriate, cleansing, disinfecting, removal of wallpaper, destroying vermin and so on. There is no right of appeal and works can be carried out in the owner or occupier's default, with costs being recovered from them.

Two notices may be relevant, allowing a reasonable time for compliance:

- Section 83 notice, for cleansing of premises.
- Section 84 notice, for cleansing and/or destruction of articles.

There is no appeal against notices. The only exception is where gassing is required (when appeal must be within 7 days), in which case rehousing must be provided (also for neighbours if necessary) for the identified duration at the local authority's expense. The Public Health Act 1961 prohibits the sale of household items known to be verminous, which should be disinfected or destroyed at the earliest opportunity.

For a flow chart on the procedure, see Bassett (1998: FC10).

Arranging the clean-up and its aftermath

Getting the premises cleansed is the priority, requiring a firm but sympathetic approach. Each and every case is different, and some examples are illustrated in Boxes 4.5 and 4.6. Early liaison with other services is necessary if there is likely to be an on-going role for the environmental health section. All involved should take sensible health and safety precautions when in the premises. Officers should be provided with protective (disposable) over-clothing including suitable shoes and latex gloves, a hepatitis B inoculation and hand disinfectant. Like many housing cases, there may be potential danger from the occupant due to mental health problems in some cases or simply because of them not wanting enforcement officers there.

Experience suggests that it is unlikely that an owner or occupier will comply with the notice(s) so works in default are usually required under section 290, with reasonable cost recovery, which can be challenged. The role of specialist contractors is paramount, and trustworthy professionals are required to carry out what can be an extremely unpleasant job. There is also the option of prosecution, but this is unlikely and even unhelpful in the circumstances.

The issue of what happens during and following enforcement action is also important. In some cases, the already hospitalised occupier dies during the works or soon after. In other cases, many possessions would have been removed and destroyed during the clearance, leaving a

Box 4.5 Theory into practice – insanitary premises and housing support need

Mr Ali's situation came to light with a tearful telephone call from his brother, who was initially too distraught to explain the situation. He had done his best to look after his brother, who had gone to pieces when their mother whom he lived with had died, some years previously. Mr Ali remained in the family home and had failed to respond to water authority demands for outstanding bills, and so the water had been disconnected for about 9 years, with no one but Mr Ali knowing.

The brother had religiously delivered bottled water and food at least once a week, but was never allowed in. Mr Ali had a mobile phone, microwave and cable television that his brother had brought him, so he maintained some contact with the outside world, but never seemed to venture out himself. The situation came to light when Mr Ali collapsed so was not contactable – his brother arranged for the police to break in, in his presence. This was the first time in years that he had been inside, to total shock, shame and disbelief at the conditions in the house he had grown up in. The doctor advised that they contact the environmental health department.

The EHO had of course seen this type of situation many times before and so was able to reassure the brother that the house could, and would, be cleared by specialist contractors, but it would first be necessary to serve a Public Health Act 'filthy and verminous' notice on Mr Ali, who would have to finance the cost of works. Owing to the lack of water supply, both toilets were completely blocked, and the bath was almost at capacity. Access was not possible into all rooms due to the sheer volume of refuse,

so 'guestimate' notices had to be served. It was clear from the smell that there was a mouse infestation, so the pest control officer immediately baited to prevent a spread of mice to neighbouring premises before the clearance got underway.

Mr Ali was meanwhile in hospital, so a large proportion of the officer's work was to reassure the distressed family and discuss a way forward with social services, who were by now involved. The EHO drafted a report on conditions on the insanitary state of the premises, stating that it was not advisable for Mr Ali to return before the clearance was done, the water supply reinstated and a full survey carried out when access became possible to determine other necessary works. The brother took the notice and letter seeking permission to carry out the works immediately following the notice to Mr Ali, who agreed to the clearance taking place, provided that the contractors could find the deeds to the property which were somewhere in the house, along with several thousand pounds.

The clearance took longer than expected as there was more refuse than anticipated. The specialist contractors took over a week to go through all items carefully. They successfully located the deeds, as well as many thousands of pounds in cash. The clearance took nine large skips. The EHO discussed the situation with the water authority who agreed, given the circumstances, to reconnect the supply for free. This was done under close supervision in case the pipework had become porous.

With the house cleared, secondary problems became apparent. Wooden doors and windows had rotted behind the piles of refuse, leaving the house open to the elements. Mr Ali remained in hospital, and it became clear that he could not return to his home. The EHO drafted a further report on the current situation to his appointed social worker. Mr Ali reluctantly agreed that he would move into a residential care home, and his application was prioritised in the circumstances. The EHO assisted the still distraught family in arranging some repair works at their expense to make the property secure. A charge arising from the clearance works on the notice was placed on the property, to reduce current stress to the family, and this was recovered by the council when the property was sold, shortly after.

Box 4.6 Theory into practice – ongoing insanitary premises

Mrs Smith was a council tenant receiving income support who had a history of mental health problems, and the environmental health section regularly visited to arrange for her home to be cleared under the Public Health Act provisions. On one occasion, it seemed that Mrs Smith had stopped taking her medication, as neighbours complained that the conditions were deteriorating again. Social services visited again and decided that she needed enforced care under the Mental Health Act, and, much to her horror, she was removed to the mental health unit for a few weeks until her condition could be stablilised.

Meanwhile, the house needed to be cleared. As Mrs Smith became more stable, she gave permission for her mental health social worker and the

EHO to visit her home. They found the house full of decaying foodstuffs still in carrier bags, stinking of stale urine and decomposing waste and with clear evidence of a major mouse problem, particularly thriving in her bedding, which was piled over with a mass of clothes, food and other waste. It was about a year since the last major clearance, and conditions had deteriorated far more rapidly this time.

The best way forward was for the EHO immediately to serve notice, arrange clearance and sort out payment from the 'dirty premises' emergency budget at a later stage. The EHO and social worker decided that it would be best to approach Mrs Smith gently in the mental health unit with the notice, explaining that the council could help with cleaning and so on. Mrs Smith, by this stage, had reverted back into a state of fury that anyone should be 'sniffing around her house' when she was not there. She grabbed the notice, tore it to pieces, and told the EHO exactly where to stick it. A few days later she was calmer and agreed in writing for the council's contractors to enforce the notice before she returned home.

Stabilised, Mrs Smith returned home and again began to accumulate refuse. Meeting after meeting followed, with various officers from social services, housing, environmental health and the voluntary sector, who tried to find a way forward. Mrs Smith did not want to move (or to be moved) and she refused regular cleaning because she thought the cleaners would steal all her possessions. In turn, the carers were either denied access or refused to do any work because they considered the conditions a health hazard. There was no viable 'in-between' option or solution available. Even progress was hampered with the enforced clearances because every night Mrs Smith would start bringing items from the skip back indoors.

The only option remaining was for the EHO to visit, serve and enforce a 'filthy and verminous' notice every 6 months to keep the worst of the refuse down and prevent further mouse infestation, so at least the situation would be kept under control.

practically empty house for the owner or occupier. The local authority or social services can sometimes provide replacements from their stores.

In an age of care in the community policies and multi-agency working, it is useful for all involved to consider the situation as a whole:

- How did the premises or person come to light and how did it get like this? What items had accumulated? Was it safe to enter? Were there any initial neighbours, friends or relatives with any involvement?
- Will the occupier continue to live there or be rehoused, with appropriate support services, perhaps in a sheltered scheme?

- Is the premises also in a poor structural condition, what role is there here – is a further notice or grant relevant?
- Are social services and/or voluntary organisations already involved or could/should they be? Is there a care plan? Is there regular help?
- Is there other help, such as from renewed contact with a family or friend?
- Is this a recurring issue, with the only realistic option to revisit and start again in 6 months?

Removal of persons in need of care

The National Assistance Act 1948 section 47 and the National Assistance (amendment) Act 1951 provide for removal of persons in need of care. This provides for those who:

- are suffering from grave or chronic illness or, being aged, infirm or physically incapacitated, are living in insanitary conditions; and
- are unable to devote themselves, and are not receiving from other persons, proper care and attention.

The proper officer must consider the interests of the person and the need to prevent injury to the health, or serious nuisance, to others and may decide that enforced care is required. The local authority has to apply to a magistrates' court for such an order, which must be jointly certified by a medical practitioner where the situation is urgent. Removal, detention and maintenance to a suitable hospital or other place may then be ordered. The order must not exceed 3 months, or 3 weeks under the accelerated procedure. Each can later be extended through the normal procedure.

For a flow chart on the procedure, see Bassett (1998: FC95).

Control of infectious disease in residential premises

The Public Health (Control of Disease) Act 1984 Part III section 31 provides for the cleansing and disinfection of premises, and the disinfection or destruction of articles to control infectious disease. This procedure applies where a proper officer has certified that the action is required to prevent the spread of infectious disease such as tuberculosis, the re-emergence of which has been associated with a rise in poverty and poor living conditions in recent years. The local authority has the discretion to pay compensation for damage arising from their action and for providing temporary accommodation whilst the disinfection is being carried out. The Act also prevents a person from letting the accommodation, whether a house or hotel, that has been occupied by persons known to have suffered from a notifiable disease, unless the accommodation has been properly disinfected. There are also controls over removal of possessions which may be infected.

This Act also gives local authorities the power to provide public cleansing (disinfection) stations to cleanse both people and articles free of charge. Very few local authorities now have these in their areas. With consent, a local authority or a county council can remove a person to a public cleansing station. If a person refuses, an order can be obtained from a magistrates' court.

Summary

- Dealing with filthy and verminous premises varies on a case-by-case basis.
- The basic remedy can be applied to the specific needs of each case, working closely with social services, the National Health Service and the housing department where appropriate, should attempt to try to find a sensitive way forward, though this can be very difficult in practice.
- There may be associated needs to be met under other legislation.
- There is specific legal remedy for dealing with infectious disease in residential premises.

4.13 Miscellaneous living accommodation

Introduction and legislation

This section is included to illustrate that high numbers of people occupy non-traditional forms of living accommodation. These fall outside the remit of the Housing Acts, but nevertheless comprise an important contribution to the nation's residential stock. However, this accommodation is distinct in that it is not generally tied to land rights. There are, therefore, all sorts of issues to contend with, not least the issue of 'trespass'.

Such accommodation is frequently linked with some of the poorest, and most 'hidden', living conditions, and remedies are sometimes inappropriate to deal with the issues faced. Where there is no recognised 'right' to live somewhere, this is usually associated with a lack of basic public health measures such as a water supply, sewer and refuse disposal. Anyone who has dealt with such issues, including trying, where appropriate, to secure alternate accommodation elsewhere, will appreciate the unique issues associated with such accommodation. Additionally, some who have housed themselves 'informally' and outside of mainstream society resent any intrusion by the State.

This section concentrates on three main types of accommodation where land rights have priority over the residence in law: self-built structures, houseboats and travellers, and each is considered in turn.

Such accommodation is covered by the following legislation:

- Public Health Act 1936.
- Environmental Protection Act 1990.
- Criminal Justice and Public Order Act 1994.

Definitions

These are incorporated as necessary in the body of the text.

Application

The basic legislation is covered for the three accommodation types below. There are marked differences between what is legally allowed and how this can sometimes be an infringement on individual liberties. Non-traditional forms of living accommodation are very much about the interface between the State versus the individual.

Self-built structures (nuisance arising from tents, vans, sheds, etc.)

The Public Health Act 1936 section 268 is concerned with living accommodation, 'the use of which, by reason of the absence of proper sanitary facilities, gives rise, whether in the site or on other land, to a nuisance or to conditions prejudicial to health'.

This is a very rarely used provision, but on some occasions local authorities are confronted with living accommodation that falls into this category. Some people house themselves in self-built structures, disused cars and so on simply because they have nowhere else to go; some may be unable to cope in more formal housing and surroundings; others may simply wish to live their life in their own way.

Local authorities have a duty to act in respect of statutory nuisance (see Section 4.4), serving notice to abate the nuisance as cited in the Public Health Act 1936 above, under the Environmental Protection Act 1990. Similar to any nuisance provisions, issues to consider when drafting the notice include responsibility for the nuisance as well as land ownership – the owner may or may not know that someone is living there, and may or may not have objections to this, possibly taking their own action. It is useful to seek advice from a solicitor, since each case will be different.

The practicalities of proposed action also need consideration, because it is likely that the occupier will have their own reasons for living in such accommodation, regardless of whether it is deemed 'lawful'. What is the occupier's point of view? Would that person be deemed statutorily homeless and eligible for social housing? Can this be tied in with the enforcement action? Is the local authority's role simply to fulfil its obligation to abate the nuisance? Such situations are rarely clear-cut and need to be dealt with on a case-by-case basis. An example of the types of issue involved when someone has chosen to live in this way is described in Box 4.7.

The Public Health (Control of Disease) Act 1984 applies to the control of infectious disease from tents, vans, sheds and similar structures used for human habitation. Local authorities can prohibit the use of such for human habitation as if the premises was a house or building so used. By-laws may be made in respect of tents, vans, sheds and similar structures used for human habitation.

Houseboats

Not all local authorities have waterways and houseboats in their area, but some areas have relatively high numbers, making a substantial contribution to local residential stock. Where there is no formal mooring, the relevant water authority may take action to move the boat on. However, in some circumstances, it can be more prudent to take a partnership approach to help tackle some of the poorest living accommodation in a local authority area so that national houseboat conditions can be gradually improved. Enforcement action seeking to improve conditions is possible regardless of mooring status, but financial assistance to improve conditions is only available in the case of where there is an established mooring – this is discussed below.

Action to tackle poor conditions on residential houseboats is shared between local authorities under the Environmental Protection Act 1990 statutory nuisance provisions (see Section 4.4) and the British Waterways Acts 1983 and 1995, enforced by British Waterways. There is very little information available about houseboats and how to deal with them because each is unique in style, construction and occupation (Stewart and Thompson 1999). People choose this lifestyle for a variety of reasons. Some do so because it is a low-cost form of owner occupation and renting. Others find it an idealised form of life, or simply wish to disappear from mainstream society.

The Boat Safety Certificate (BSC) was introduced by the British Waterways Act 1995 to improve safety standards on all boats, not just residential ones. These certificates are issued under the Boat Safety Scheme, which is jointly responsible to British Waterways and the Environment Agency and sets minimum requirements for construction and equipping of boats. Where a boat satisfies the required standards a BSC is issued which is valid for 4 years. Enforcement is by prosecution of the boat owner and/or removing the boat from the water, which may result in households becoming homeless.

Box 4.7 Theory into practice – the right way forward?

A local authority had received a complaint about a man living in the woods and 'squatting outside for the toilet', which a local resident found 'quite disgusting'. Intrigued, the EHO visited to find that this was true – Mr Sheppard, in his late 60s, had built himself a home from recycled materials, on land that belonged to someone else. His home had the appearance of an allotment shed, constructed from various wood and plastic, and Mr Sheppard was very proud of his achievement. It emerged that he had lived there for several years after being evicted from a rented caravan, without anyone in 'authority' knowing.

As ever with such situations, things are rarely as simple as they seem. It emerged that Mr Sheppard's elderly mother lived alone, only a 5-minute walk away, in a large three-bedroomed house, across the border in another local authority area. He said that he visited every day for the toilet and the bath, but would not live there because he liked to live in the woods with the birds, trees and fresh air. Besides, he said, he was a grown man and should not expect his mother to put him up.

The local authority in whose area he was living tried to encourage him to apply as homeless and offered him what it felt was more appropriate accommodation, but Mr Sheppard was not interested in a third-floor council flat – he liked it here too much, with the birds. The local authority carried out a land search to find out that three people, now elderly, had inherited the land from a long-lost relative, and had no interest in it, as it was by now enclosed by housing developments. They had no objections to Mr Sheppard living there at all, and were glad that at least someone had use of the land.

Informal action was clearly not going to change anything. The local authority was then in the difficult position of having a duty to serve notice under the Environmental Protection Act using the Public Health Act 1936 provisions on Mr Sheppard as the person responsible for the statutory nuisance. Mr Sheppard had by now had enough of various 'busy bodies' interfering in his life – he said that social services had visited to see if he was 'mad'; housing officers had harassed him into moving into flats and now the EHO was about to pull down the home he had built with his own hands and which he loved dearly. Mr Sheppard broke down and sobbed uncontrollably, asking why everyone would not just leave him alone – what harm was he doing?

It was a duty to serve notice, but not to enforce it. However, with continued complaints and lobbying, the local authority had to take matters further. It decided to enforce the notice, and given the unusual circumstances, made one final offer to Mr Sheppard of another council flat. Officers visited again to try to resolve the situation informally, but in the end this proved ineffective.

The local authority employed contractors to dismantle the accommodation in Mr Sheppard's default of the notice. They took the remains of it away for disposal at the local authority's cost. Mr Sheppard was heartbroken, but not defeated. He found himself a plot of vacant land a few minutes' walk away in the next local authority's area and built himself something new to live in.

The relevant legislation for residential standards in occupied boats is contained in the Environmental Protection Act 1990 statutory nuisance provisions (see Section 4.4). The whole condition, including amenities of the boat, should be fully considered before deciding whether conditions are prejudicial to the health of the occupants. Whilst the BSC may go part way to achieving environmental health standards, it is only concerned with the safety of existing standards on the boat and its physical conditions, not whether it is suitable as residential accommodation. The environmental health legislation goes further and incorporates amenities.

Based on recommendations by the CIEH, British Waterways has produced guidelines for residential boats (see Leaflet section page 278) that incorporate the following requirements:

- Water storage and supply.
- Sanitary accommodation.

- Facilities for the preparation and cooking of food.
- Space heating.
- Artificial lighting.
- Ventilation.
- Fire and safety precautions.
- Life saving equipment.
- Repair.

Enforcement is by service of an abatement notice under the Environmental Protection Act 1990 (also see Section 4.4) requiring relevant action. It may be difficult to establish on whom to serve the notice because the occupier may not be the owner. There are normally no formal ownership documents for boats and a written statement of ownership is normally all that is available, even on sale of the boat. If ownership cannot be formalised, it may be advisable to serve it on the occupier to both postal address and attach one to the boat. Even where a notice has been properly served, the boat may simply disappear, within or outside of the local authority's jurisdiction. Enforcement of the notice would be extremely difficult.

For a flow chart on the procedure, see Bassett (1998: FC33).

Home Repairs Assistance (HRA) is another possibility, available at the discretion of each authority, and subject to the applicant having a recognised right to moorings. There are many conditions attached that may exclude qualification. Further information is contained in Section 4.6, but in terms of houseboats, HRA is available as follows:

- For applicants over 18 years of age with an owner's interest, where the boat is their main residence.
- The applicant must have occupied the boat for at least 3 years immediately preceding the application.
- The boat must have permission to be moored in the local authority area for that period.
- The applicant must be in receipt of an income-related benefit and have a duty or power to carry out works.

Some of these criteria can be difficult to prove, and many travelling people would fall outside of them.

Another possibility is the Department of Social Security (DSS) Social Fund. This may be available should a boat need repair or adaptation to make it safe or if minor repairs are necessary. This can be in the form of a community care grant, budgeting loan or crisis loan. Only those on income-related benefit are eligible. It does help the applicant to gain support from the environmental health department among other agencies on application. The DSS then decides on the priority and whether a loan or grant will be given and the amount.

There are many practical issues to be addressed by enforcement authorities in dealing with houseboats. The main problem is that since houseboats can be mobile, enforcement can be extremely difficult and authorities may not be fully aware of the nature and extent of residential houseboats in their area. British Waterways may have this information, but with prosecution and taking control of the boat being its only course of action, it acting alone may not result in a viable outcome, particularly where the boat is occupied. Occupation of houseboats may be 'hidden' from British Waterways due to cost implications in terms of licence requirements.

From an environmental health and rehousing point of view, the boat owner may be willing to provide information on occupancy and conditions as well as access, since this may result in a more viable form of assistance than the British Waterways is able to provide. Subject to the Data Protection Act 1998, British Waterways can provide information about the 'legality' of the boat on the waterway and the patrol officer is well placed to provide up-to-date information as to the boat's location. A joint database of boats, including a photograph, registration number and name, could help track problem boats.

A houseboat may pass the BSC but remain a statutory nuisance due to inadequate amenities to make it habitable. Examples include where there is no gas or electrical installation, which would not be included in the BSC inspection simply because of not being there at the time of the inspection. Local authorities would, however, require that some form of artificial lighting be provided. If local authority enforcement or grant-aid resulted in an electrical installation being provided, a further BSC inspection would be immediately required on the additional item(s). It may be more economic for a local authority to upgrade a boat with grant-aid than to face rehousing obligations, and experience suggests that boat owners normally choose to remain on the boat.

In the event that the boat is in such poor condition that it has to be removed from the water, there may be rehousing implications. This would normally follow the homeless route, and each case would have to be individually assessed. In view of the unique circumstances surrounding residency, boat owner-occupiers may be accepted as homeless, provided that other relevant conditions for the homeless application are also met. The local authority may not always have an obligation to rehouse. If the occupiers are rehoused due to the boat's conditions, it would be prudent to work jointly with British Waterways to ensure that the boat is either removed from circulation or improved to required standards.

Conditions on houseboats can be amongst the worst, and most dangerous, residential conditions facing local authorities, and attempting to apply the legislation to some houseboats can be extremely complex and far from clear-cut. An example, with illustrations, is shown in Box 4.8.

Box 4.8 Theory into practice – addressing the apparently impossible

A local authority had developed a strategy to tackle the many houseboats moored both formally and informally in its area. It was aware from previous complaints and proactive survey work that many were in severe disrepair and lacked basic amenities. Some, in particular, looked even dangerous from the tow-path.

Aware that the local authority might be able to assist her, one woman came into the council offices with her two young children. She said that she had bought her boat from a man she met for a thousand pounds and had no money to fix it, but had started some works. She had a Calor gas supply, but no toilet or sink, and the bilge pump had broken so she could not pump out the excess water, so the boat had almost sank twice. She wanted to have a party for her little girl's birthday, but could not with the boat as it was. She said that her partner sometimes stayed and helped, but she found it difficult to get by on her benefits, and she could not get a mooring. It was immediately clear, without even visiting the boat, and seeing both her and the children, that the situation was going to be extremely difficult to resolve.

Officers visited and found that the boat was in a potentially dangerous situation, not on a formal mooring, and looking likely either to sink or explode, or perhaps both. The question of what to do next was difficult. The woman wanted to stay on the boat. She was not eligible for home repairs assistance because the boat was not on a mooring, and it was very unlikely that she would receive assistance from the DSS social fund. In any event, this would go little way toward the major works that would be needed even to make the houseboat safe. The housing department tried to find her a house to rent, but she would not live in formal housing because of the 'constant ringing in her ears'. They tried to find her a houseboat to rent, but there was nothing locally.

Social services got in touch, requesting a report on the living conditions. They were very concerned about the children living there. The EHO drafted a report based on British Waterways and the CIEH's guidance, outlining that conditions were potentially dangerous, the houseboat was in serious disrepair with no apparent remedy, there were no amenities aboard, she had no formal mooring so was ineligible for assistance, etc. The partner, meanwhile, returned and beat her up because she had been in touch with the council, and he did not want the children to be taken away.

The situation ironically resolved itself, as some of these situations do, when the children were taken into care and the mother moved the houseboat overnight to an unknown location along the canal or a local river, which sank without trace. She then fled her violent partner to an unknown location, apparently finding somewhere else to live. In such cases, it is difficult to know whether the public services' role was of any help – but if there had been no intervention, the boat may have sank overnight with a woman and her two children still sleeping on board.

There are also specific regulations for canal boats. A canal boat is defined as including any vessel, however propelled, used for the conveyance of goods along a canal, not being:

- a sailing barge that belongs to the class generally known as 'Thames Sailing Barge' and is registered under the Merchant Shipping Act 1874–1928 either in the Port of London or elsewhere; or

- a sea-going vessel; or
- a vessel used for pleasure purposes only; or
- a vessel carrying a cargo of petroleum.

The Public Health (Control of Disease) Act 1984 provides the Secretary of State a duty to make regulations:

- fix standards for space and occupancy, ventilation, provision for separation of the sexes, general healthiness and convenience of occupation;
- promote cleanliness and habitable condition; and
- prevent the spread of infectious disease on canal boats.

The most recent regulations are the Canal Boat Regulations 1878 (as amended in 1925 and 1931). Those parts of the regulations still in force are now quite dated and the definition may no longer apply to many boats in their current use. The extent to which they are enforced by local authorities is unclear, although local authorities are empowered to enter and inspect canal boats.

Gypsies and travellers – the need for sites

There remains much controversy as to gypsies and travellers' lifestyles, perhaps because as a group they are generally more visible to the general public than the above. Gypsies and travellers comprise distinct groups, but are often combined for policy purposes. This section looks at definitions, before turning to relevant legislation that addresses land rights, living accommodation and formal sites in respect of mobile communities.

Gypsy – a person of nomadic habitat, whatever race or origin, but the term does not include members of an organised group of travelling showmen or persons engaged in travelling circuses, travelling together as such (Caravan Sites and Control of Development Act 1960, as amended by Criminal Justice and Public Order Act 1994). The term is generally accepted to incorporate those who move from place to place for their livelihood.

New Age Traveller – a term that emerged in the 1990s, but it has no legal definition, and it mainly comprises those who have left settled communities.

The Caravan Sites Act 1968 provided a duty for County Councils, London Boroughs and Metropolitan Boroughs to provide sites, with 100 per cent capital grant from 1978. More than 300 gypsy sites were provided, which cater for around 6,000 caravans – about half of all the gypsy caravans in England. There are also privately owned sites, and the combination of these means that around 80 per cent of gypsy caravans occupy authorised sites. This also means that 20 per cent are not, and they cannot stay on sites. There has been a rapid increase in gypsy families, from 4,750 caravans in 1965 to 13,500 by 1993 (prior to the new Act), but only 38 per cent of local authorities had satisfied their duty to provide sites. The Secretary of State had the power to require sites, but there is little evidence that this happened. These figures show that around two-thirds of English local authorities failed to provide sites, despite the government financing the full cost of capital works (Ormandy 1999). The lack of sites with growing numbers of caravans inevitably means that many have no designated place to stay. This also meant that many have no immediate access to basic public heath measures, such as drinking water, toilets and refuse disposal.

In spite of growing numbers of travelling communities with nowhere to stay, the Criminal Justice and Public Order Act 1994 repealed the local authority duty under to provide sites for gypsies (different from caravan sites discussed in Section 4.14) and control unauthorised encampments. The Act:

- removed the capital grant to cover site provision;
- strengthened powers to deal with gypsies on unauthorised land, with trespass becoming a criminal offence, and direct unauthorised campers to leave;
- gave the police new powers to remove trespassers, seize vehicles and limit/prevent raves;
- allowed for tolerance where unauthorised camping was not causing a nuisance; and
- encouraged local authorities to consider gypsy sites as part of their local plans.

The then DoE (1994) recognised that not all unauthorised encampments caused nuisance and moving people on may cause a problem elsewhere. The government suggested that local authorities 'tolerate' some sites, and identify some stopping places near to regular stopping places, with the provision of basic amenities. The government has more recently indicated that local authorities should locally determine action to take in respect of gypsies and travelling communities through developing planning strategies that are realistic and meet need. This requires a flexible balance of travellers' and settled communities' needs and aspirations according to local circumstances – frequently an emotionally charged debate. The guidance for local authorities and the police draws together experience on good practice to encourage partnership working between local authorities, the police, health service agencies, education departments and so on to help manage unauthorised camping.

The legislation has had little impact on numbers of gypsies and travellers who are counted twice per year by local authority environmental health departments on behalf of the DETR. Government statistics, cited in Ormandy (1999), show that:

- in January 1998 the number of local authority gypsy sites in England was 329 (5,265 pitches); and
- in July 1998, a count of gypsy caravans in England recorded 3,700 unauthorised encampments, 5,997 on authorised sites and 4,210 on unauthorised private sites. This amounts to 13,545, mainly due to an increase of private sites.

Local authorities are now required to develop local strategies to deal with non-settled communities by taking into account:

- local planning policies and realistic planning criteria allowing for an overall strategy for travellers and gypsies, including a needs assessment, site and service provision and identification of temporary stopping places;
- the views of the local authority and local communities;
- the need for flexible local services; and
- good practice guidelines, such as from the DETR and Home Office 1998, to promote better joint working between local authorities, the police (for eviction policies) and other services.

The government is keen for local authorities and the police to develop strategies to deal with travelling communities and unauthorised encampments in their area. Considerable time and resources are spent in effecting evictions and clearing land, which does not resolve travelling people's need for somewhere to stay. For this reason, they have developed good practice guidance encouraging authorities to develop local strategies to deal with travelling communities and manage unauthorised encampments (DoE 1994).

For a flow chart on the procedure, see Bassett (1998: FC105).

Summary

- Issues here illustrate how people's living accommodation is closely allied to land rights – and land rights normally take precedence over living accommodation standards and public health measures.
- No one is legally allowed to live in a self-built structure and living in houseboats and caravans are only legally permitted on recognised moorings or formal sites.
- Where there are no legally recognised rights to land, eviction normally follows.

- There is little ability for people to occupy 'non-traditional' forms of housing, even where it is at no cost to the local authority (e.g. through loss of local authority housing unit).

Acknowledgement

A more detailed version of the houseboat section first appeared in the CIEH's official journal, *Environmental Health Journal*, 107(5).

4.14 Caravan site licensing and control

Introduction and legislation

Many local authorities have sites that are lived in on a permanent residential basis. Some are becoming increasingly desirable as previous owner-occupiers release their equity in formal housing and move into relatively low-cost mobile homes on a licensed site. Many site managers are moving into this type of site ownership. Other sites are less desirable and occupied or rented by people unable to afford alternate forms of housing. Dealing with caravan sites involves enforcing Model Conditions under the Caravan Sites and Control of Development Act 1960. This Act is only concerned with residential sites and does not refer to licensing of camping sites.

Definitions

Caravan means any structure designed or adapted for human habitation capable of being moved (e.g. towed) and any motor vehicle so designed or adapted. It does not include rolling stock or tents or a structure that cannot be moved on a highway once it has been assembled on site due to its dimensions, where the length is 18 metres, width 6 metres and height 3 metres.

Caravan site means land upon which a caravan is stationed for the purposes of human habitation and land used in conjunction with it.

Occupier means, in relation to any land, the person who, by virtue of an estate or interest held, is entitled to possession or would be so entitled but for the rights of any other person under licence granted in respect of the land.

Application

The occupier of land used as a caravan site must have a site licence, but the following exemptions apply:

- Incidental use in the curtilage of a dwelling-house.

- Single caravan used for no more than two nights, and 28 days in any 12 months.
- Holdings of 5 or more acres, if not occupied for more than 28 days in 12 months and with a maximum of three caravans.
- Sites exempted by the minister.
- Sites for exempted organisations for up to five caravans.
- Meetings organised by exempted organisations.
- Sites for agriculture and forestry workers; building and engineering services; travelling showmen.
- Sites occupied by a local authority.
- Gypsy sites occupied by county councils (see Section 4.13).

Applications, licensing and model conditions

The occupier of land must apply to the local authority in writing and there is no fee payable. The local authority must issue a licence within 2 months (or longer if agreed) unless:

- no formal planning permission has been granted;
- the applicant has had a licence revoked in the previous 3 years; and
- planning permission will expire within 6 months of date of application.

For a flow chart on the procedure, see Bassett (1998: FC88).

The local authority may attach conditions to the licence and must have regard to the Model Conditions issued by the Minister from time to time, most recently in 1989 (DoE 1989). A summary of these conditions appears in Table 4.8. Works to comply with these conditions can be required in a specified time and works in default procedures apply. The local authority must consult with the fire authority before issuing a site licence, or when varying conditions.

The Audit Commission (1991) recommends that local authorities inspect caravan sites within their area on an annual basis to ensure compliance with the licence conditions.

Power of entry

An authorised officer, after 24 hours' written notice and producing an authenticated authority document, can enter any land used as a caravan site to:

- determine conditions attached to licence;
- ascertain contraventions; and

Table 4.8 Current DoE caravan site model conditions

Heading	*Requirements*
Site boundary	• To be clearly marked • Local authority to have a plan of the layout • Three metres clear inside the boundary
Density and space between caravans	• Caravans to be not less than 6 metres from another (where occupied) • Caravans not to be less than 2 metres from roads • Porches protrude no more than 1 metre into 6 metres, or to be of open type • Awnings to be 3 metres apart, with no sleeping accommodation, and not facing • Eaves, drainpipes, bay windows not to reduce the distance between caravans to less than 5.25 metres • Garage, shed or covered store to be non-combustible, with windows not facing units so as not to prejudice a fire escape • Density to not exceed 50 units per hectare (excluding lakes, roads, etc.)
Roads and hardstanding	• No caravan to be more than 50 metres from a road • Gateways to be a minimum of 3.1 metres wide and 3.7 metres high • To be suitably lit • Emergency routes to be kept clear • Hardstanding to be provided under a caravan, extending to allow safe access
Fire and firefighting appliances	• Fire points to be marked and weatherproofed, with no caravan more than 30 metres away • Water standpipe to be provided with pressure sufficient for a 5-metre jet • Reel to meet BS 5306 Part 1; to be 30 metres long and housed in a red box marked 'Hosereel', or water fire extinguishers with a water tank of 500 litres, and provided with two buckets • Fire alarm, inspection log book and fire notice to be provided • Fire hazards to be notified • Telephone or notice with an address to be provided
Liquid petroleum gas (LPG)	• Bottles to be kept in a separation area • All installations to comply with Guidance Notes
Electrical installations	• Sufficient supply to be maintained and inspected annually and certificated; overhead lines to have warning display
Water supply and drainage	• To be provided with a suitable water supply and provision be made for foul and surface water drainage
Refuse	• Bins, and their emptying, to be provided
Parking	• One car only between caravans
Recreational space	• One-tenth of the area to be provided for recreation space – not normally applicable for sites with one to three units
Notices	• Site licence to be displayed • Fire plan, with contact name and numbers for fire, police, ambulance, doctor and location of nearest telephone, to be provided

Source: based on DoE (1989)

- ascertain if the local authority should carry out any works, and do so if necessary.

If entry is refused, the local authority can apply for a warrant to enter.

Caravan Sites Act 1968 and Mobile Homes Act 1983

Relations between caravans (more commonly now referred to as mobile homes or park homes) and site owners are largely governed by the Caravan Sites Act 1968 and the Mobile Homes Act 1983, which updated many existing provisions. The Mobile Homes Act applies to private and local authority sites. It introduces new rules governing ownership (but not tenancies) in respect of mobile homes, including security of tenure, sale of mobile homes and agreements between site owners and residents. The act requires that site owners give residents a written statement about their rights in a specified format on matters such as pitch fees, charges and obligations on residents.

Proposals for change

There remain some concerns as to the awareness and security of ownership rights and the scope and effectiveness of existing enforcement regimes, including alleged over-regulation. In view of the concerns, the government established a Park Homes Working Party in 1998 to consider existing controls and proposals for change to achieve a viable balance between those involved. The Working Party's report was published by the DETR in July 2000, and made several key recommendations, including providing guidance material, amendments to current Model Standards and the introduction of good practice codes on harassment and site licensing. At the time of writing, further research continues.

Summary

- Many local authorities have residential caravan sites within their area that require licensing.
- Annual inspection is recommended to ensure compliance with model conditions.

4.15 Home energy conservation acts

Introduction and legislation

A background to fuel poverty, energy efficiency and conservation was set out in Section 2.7 and should be read in conjunction with this section.

International agreement at the Rio Earth Summit sought to reduce 'greenhouse gases' to 1990 levels by 2000, which included a commitment to reduce domestic energy consumption and emissions. Housing's role is national, regional and local, with the practical implementation of policy based on local authorities as strategic enablers and coordinators of home energy conservation legislation.

Local authorities have, to some extent, always played a role in domestic energy in their housing renewal strategies, and the new energy Acts sought to consolidate and build on this. This section provides a brief overview of the Energy Conservation Act 1995 and the Energy Conservation Act 1996, which govern domestic energy efficiency. Department of the Environment Circulars 2/96 and 5/97 support these Acts respectively. This section also reviews the Home Energy Efficiency Scheme (HEES).

Definitions

These are incorporated in the body of this section as necessary.

Application

The Home Energy Conservation Act 1995 (HECA) provided a framework for local authorities to review strategy in respect of domestic energy conservation. The Energy Conservation Act 1996 (ECA) extended the provisions to HMOs and some houseboats. Through the Home Energy Conservation Act, each local authority became an Energy Conservation Authority required to prepare an energy conservation report outlining their proposals to reduce domestic energy consumption by 30 per cent.

The report had to incorporate the following key features:

- An assessment of cost of measures.
- An estimate of decrease in carbon dioxide emissions.
- A statement of policy for taking personal circumstances into account (e.g. including existing strategies).

In addition, local authorities had discretion to include:

- an estimate in reduction of nitrogen oxides and sulphur dioxides;
- the possible creation of jobs;
- an assessment of savings in fuel bills; and
- other factors.

Implementing HECA (DoE 2/96) and ECA (DoE 5/97)

Local authorities were seen as catalysts for change through developing new approaches to encourage domestic energy efficiency. Their role was seen to encourage physical changes in dwellings, through the availability of grants and loans, providing information, advice, education and energy promotion and to be imaginative in forging partnerships with new agencies. ECA extended the definition of residential accommodation from the housing legislation definition, plus certain mobile homes, to incorporate houses in multiple occupation and some houseboats.

In line with much other legislation, there was to be no additional government funding. There was also no power under this legislation to require works to be done. Local authorities would have to make best use of existing resources, including grants and cold weather payments, as well as developing new partnerships with the private sector, including electricity and gas suppliers. The scope for adding impetus to anti-poverty strategies and energy efficiency strategies was clear.

Proposals for change

In view of continuing pressure to improve domestic heating and energy efficiency, the Warm Homes and Energy Conservation Bill was presented to the House of Commons in 1999. This Bill requires the Secretary of State to develop a programme of action and to facilitate carrying out a comprehensive package of home insulation and other energy efficiency measures to at least 500,000 households per annum over 15 years.

The comprehensive package of home insulation and other energy efficiency measures includes:

- Cavity wall insulation.
- Loft insulation.
- Under-floor insulation.
- Insulating hot water pipes.
- Insulating pipes or other plumbing facilities.
- Draught proofing.
- Controlling domestic heating systems.
- External and internal wall cladding.
- Providing low emissivity glazing.

It remains to be seen how the Act will be finally worded and resourced.

Home Energy Efficiency Scheme (HEES)

The Energy Action Grants Agency introduced the Department of Energy's Home Energy Efficiency Scheme (HEES) in 1991, established under the Social Security Act 1990. The scheme has a wide insulation remit, including giving grants and advice about loft insulation, tank and pipe insulation, draft-proofing, space and water heating for across all tenures. The criteria for eligibility change from time to time, but in general the grant is available to those on income-related benefits. The scheme is not administered by local authorities but forms an important part in their local housing strategy alongside other home energy conservation issues. HEES is an advantage to local authorities in that they can access non-local authority resources in respect of their housing stock, operating alongside larger-scale renovation projects across all sectors, individual enforcement or grant-led work, and energy-efficiency works under home repair assistance.

Summary

- HECA and ECA were introduced to reduce domestic energy consumption.
- Much is already being done by local authorities, but there is always scope for increased action.
- No additional funding was made available, or specific legal power to require works, so local authorities, as 'energy conservation authorities' have had to develop new partnerships and develop educative programmes.
- The legislation adds impetus to tackling fuel poverty, but further resources are required.

4.16 Reconnection of services

Introduction and legislation

Sometimes landlords fail to pay bills for gas, water and electricity, leading to supplies being cut off and tenants being left without an essential service. Tenants can be powerless to get the landlord to pay the bill so that the supply can be reconnected without delay. Tenants left without gas, water or electricity need it reinstated within a working day. The quickest route to arrange this is through the Local Government (Miscellaneous Provisions) Act 1976 (as amended) section 33.

Definitions

None specific.

Application

The Local Government (Miscellaneous Provisions) Act 1976 (as amended) section 33 can be used to restore gas, electricity and water services where they have been, or are likely to be, disconnected through the landlord's default of payment.

Notification of disconnection may come from either a tenant or the relevant service authority, but the local authority has to have a written request for reconnection from a tenant before making reconnection arrangements. This can be either to restore the supply or to prevent it from being disconnected where this results from the landlord's failure to pay a bill. The amount of the bill, plus the local authority costs, can be recovered from the landlord and a charge placed on the property. An example of such a situation is illustrated in Box 4.9.

Local authorities need to have sound procedures in place both to arrange the reconnection, pay the bill and recover the money from the landlord. Sometimes the landlord will arrange the reconnection following a telephone call from the local authority outlining the case that it would be easier and cheaper to pay the bill immediately without incurring further charges.

If this approach were not successful, the local authority would need to arrange the reconnection on a works in default-type basis. In practice, the service provider would need a request and promise to pay the outstanding bill and reconnection fee (if appropriate) in writing from the local authority, normally by facsimile on headed paper. They would then arrange the reconnection of the service at the standard fee, normally on the same day if possible and invoice the local authority accordingly. The local authority would make payment and seek reimbursement from the landlord through normal financial channels. Failure to pay would result in a land charge being placed on the premises, with interest accumulating until repaid in full.

Sometimes the fact that a landlord has failed to pay a bill may point to wider problems with a rented property. Notification from the service provider, or a call from a tenant that a service is to be, or has been disconnected, may bring the premises to the attention of the local authority for the first time, and wider enforcement works may also prove necessary.

Summary

- Local authorities can prevent disconnection or promptly rearrange for reconnection of essential services, and they require the cost back from the landlord.

Box 4.9 Theory into practice – reconnection of water supply

A local authority received notification by fax from a water authority that it had disconnected the water supply in a local residential premises due to non-payment. The environmental health section of the local authority was not previously aware that the property was rented, although it later emerged that it was let as an HMO, mostly to single people on housing benefit.

On visiting, the EHO found two people there, who said that they could not flush the toilet or make themselves a cup of tea because they had had no water since last night. The EHO encouraged one of them to put something in writing, but the tenant was wary of doing so for fear of being evicted and causing trouble with the landlord and the other tenants. Eventually, the tenant wrote a couple of lines on a paper bag and scribbled a signature. The EHO could then return to the office to make arrangements for the reconnection.

It took most of the day to telephone the water authority, fax a confirmation that the local authority would pay the bill in the interim, and ensure that someone came to reconnect the supply at the earliest opportunity. The tenant would give only few details about the landlord, so the details were confirmed with the housing benefit (whose records illustrated a rapid tenant turnover) and council tax departments (where there was a history of late payments), and a section 16 and covering letter drafted for the longer-term default payment. A land registry search was also instigated, as an early response to the section 16 was not expected. A charge was placed on the property to cover the cost of service reconnection that the council had paid.

The EHO then had to make other arrangements in respect of the HMO; there were no means of escape, many repairs needed doing, and there were many management issues to address. It later emerged that the house was in the process of being repossessed. All the tenants left, a pipe burst in the loft and the house was flooded, so the EHO ironically had to arrange for the water to be disconnected, as there was no one else to do so. Someone broke in at the back and pulled down some of the damaged ceilings. Unable to do much more at this stage, the house had to be boarded up to prevent further vandalism.

Several local people, having seen the property boarded up, and interested in purchasing it, telephoned the council to ask about it. With permission, their details were forwarded to the bank who had repossessed it – it was keen to sell the house at the earliest opportunity.

The bank, as current owner, forwarded a cheque for several hundred pounds to reimburse the council for paying the outstanding water bill and reconnection fee so that the land charge against the property could be erased. The property was eventually sold by the bank and completely renovated, and all outstanding notices were cancelled.

4.17 Disposal of dead bodies

Introduction and legislation

More people are dying alone with no friends or family to arrange their burial. Where there are no relations or friends to arrange a burial, the practical role of doing so falls to the local authority Environmental Health Department under the provisions of the Public Health (Control of Disease) Act 1984 sections 46–8.

Definitions

None specific.

Application

Local authorities have a statutory duty to arrange the burial or cremation of a person who has died or been found dead in their area, where there is no one else to do so. They may arrange a cremation or burial subject to the wishes, or likely wishes, of the deceased. The local authority can recover the cost of the burial from the estate or any person liable to maintain the deceased person immediately before their death for the purposes of the National Assistance Act 1948, recoverable as a civil debt. In some cases there is no estate, in which case the local authority would have to finance the burial or cremation.

In practice, the local authority would also need to make all arrangements for the estate through the wishes expressed in the will, if a will were found. Where no relatives can be traced, any remaining funds from the estate would normally revert to the local authority general fund.

In some cases, the local authority may also find that the premises in which the person has died is insanitary, as discussed in Section 4.12. In such a case, the local authority would additionally have to arrange for the premises to be cleared and charged for as appropriate.

The Public Health (Control of Disease) Act 1984 also provides for the removal of dead bodies to the mortuary or for immediate burial on certificate of the proper officer, where retention of that body would endanger others. The local authority would have to finance this and arrange burial within the time specified.

Summary

- Local authorities have a statutory duty to arrange for disposal of dead bodies where there is no one else identified to do so.

Chapter 5

Private sector housing strategies

Chapter 5 considers housing strategies and looks at the changing nature of such strategies and their current role and function in a rapidly changing environment. It focuses on government requirements in developing and delivering comprehensive housing strategies and offers a general overview of their increasing importance in terms of delivering far more than just housing. It looks at some key issues involved in developing housing strategies before considering establishing housing strategies to tackle houses in multiple occupation and making best use of the private rented sector. It provides an overview of urban regeneration – the reasons for decline and how successful strategies need to act upon these, before looking at sources of funding for regeneration projects. It concludes with a summary of current concepts in regeneration – community empowerment, partnership working and social exclusion as key features to be addressed in evolving housing strategies.

This chapter is presented as follows:

5.1 Changing nature of housing strategies
5.2 Key features in developing a strategy
5.3 Developing a strategy for HMOs
5.4 Developing a strategy for vacant dwellings and optimising the private rented sector
5.5 Urban regeneration
5.6 Issues in regeneration – participation, partnerships and exclusion to inclusion

5.1 Changing nature of housing strategies

Outline

The need for well-thought-through strategies has been a result of massive social, political, economic and legal changes, both to housing and in wider society. The 1991 Audit Commission report (Audit

Commission 1991) criticised local authorities for their general lack of strategies to make best use of available resources in private sector stock. In 1992, it published a further document advising local authorities on how to develop housing strategies (Audit Commission 1992). In addition, the general drive toward quality assurance through the 1980s and 1990s has added impetus to the need for written strategies and accountability in service delivery.

There has been a change in emphasis from a 'bricks and mortar' approach to housing and a return to individual responsibility for housing conditions, with reduced state involvement and resource. Changing ideologies over the past 150 years have been reflected in housing policy and the role the State should take in people's lives. More recently, housing strategies have become as much about social exclusion, unemployment and training as about housing itself. What is now key is involving local communities in local employment and training schemes, forging new partnerships across sectors, and this is perhaps where the future of housing strategies will increasingly lie. This section defines housing strategy, before looking at its traditional and evolving purposes. Partnerships have become increasingly important in recent years and the term is defined in Section 5.6.

Changing nature of housing strategies

General introduction

The term 'strategy' is a military one. It is about how to win a war. In terms of housing, strategy can be defined as winning the war against poor and unsuitable housing. Increasingly, though, housing strategies have taken on a wider context than their traditional bricks and mortar approach. Local housing strategies are now much more about the bigger picture, and bringing wider socio-economic change to communities. The terms 'policy' and 'strategy' are frequently interchanged. In general, however, policy refers to what is going to be done, whilst strategy is about how this is going to be achieved (Figure 5.1).

Current nature of housing strategies

Local housing strategies have a wider remit than just responding to an annual government requirement, and this is discussed below. Well-designed strategies are crucial in a climate of decreasing capital expenditure; they require that local authorities review what they are doing to ensure best use of available resources, as well as the possibility of attracting other possible funding in providing services. Increasingly, the emphasis is on prioritising action to areas and conditions of greatest

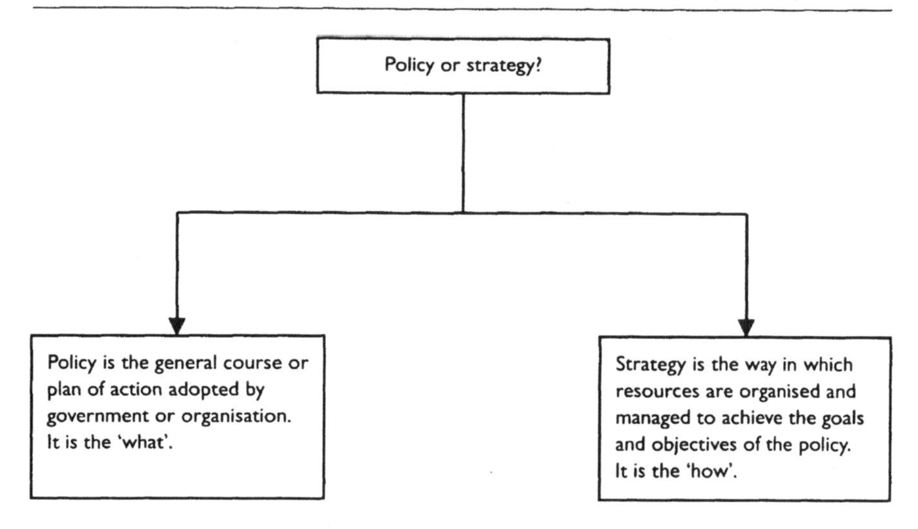

Figure 5.1 Policy and strategy

need, rather than by being reactive. Good strategies ensure that action is coordinated across the local authority and that housing functions are reviewed, to avoid duplication and bureaucracy. They increasingly have a wider remit than just housing functions.

Since strategies are now more diverse than providing social housing, or enforcing conditions in the private sector, they can take increasingly varied and specialist forms. A strategy may be:

- Area based.
- Based on property type or condition.
- Issue based.
- Client based.
- Existing – including the local authority's own stock, SRB projects, estate action.
- Other.

The way such strategies might fit into local authority organisational structures, and at which level, are illustrated in Figure 5.2.

Goss and Blackaby (1998) summarise the purpose and importance of housing strategy as including the following features:

- To assess the current and future balance of local housing supply and demand – the features of the housing market, the impact of conditions on people's lives and the achievement of related social goals.

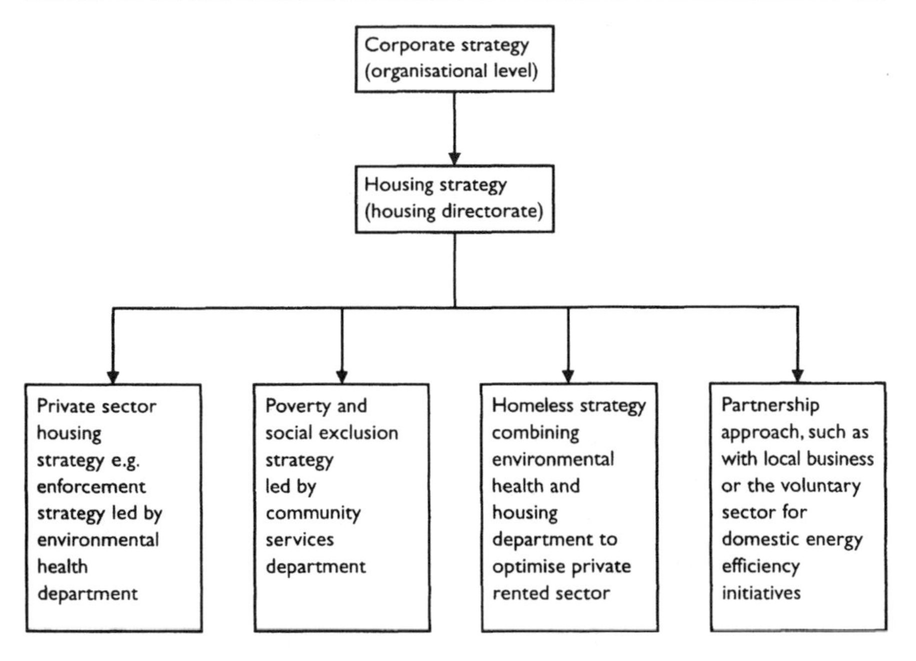

Figure 5.2 Example of how a housing strategy might operate

- To guide local action and to plan effective intervention.
- To set a direction for managerial action for the local authority and for other agencies in addressing local housing action.

Government requirements

Local authorities are required to submit an annual housing strategy to the Department of the Environment, Transport and the Regions. This strategy document needs to include:

- Description of general housing policy.
- Major housing issues in the area.
- Overall approach to local housing.
- Particular issues to be addressed.

The DETR has issued guidance about the submission, with increasing emphasis being placed on performance over time. The document forms the basis of the HIP allocation. The strategy must be reviewed annually to keep it up to date. It needs to be planned, programmed, regularly reviewed and monitored (Oxby 1999).

The emphasis of local housing strategies has increasingly moved toward the private sector in recent years. This requires greater input not only from environmental health officers, but also from housing advice and homeless officers, whose remit increasingly lies in that sector. With a decline in traditional social housing, and with growing numbers being accepted as homeless, the private sector is seen to meet increased demand. There has been increased government guidance on developing proactive strategy, and documents such as the Audit Commission reports (1991, 1992) and DoE guidance on developing private sector housing strategies (DoE 1996) in particular, encourage:

- increasingly integrated and planned area-based action;
- greater use of low-cost private sector accommodation, backed up by good enforcement;
- bringing vacant properties back into use;
- increased use of housing advice officers to help prevent homelessness, harassment and eviction;
- greater sensitivity toward special needs groups, including people with disabilities and the elderly; and
- an increase in multi-agency working with statutory and non-statutory organisations involved in housing and social services.

Housing strategies essentially need to meet local housing need, however that is defined. Local authorities have moved from being mainstream providers to an emerging enabling function with an increased emphasis on local strategies. There is no national housing strategy as such, but rather a set of objectives to be met in housing service delivery. Part of the evolving process has also increased the role of agencies and a sense of 'competition' in providing traditional local authority housing functions. There has been increased contracting and the development of partnerships and commissioning across many fields. There has even been a shift to the private sector in providing new forms of assistance, including home services and building societies. What is needed is to stand back and take a real and honest look at what is actually going on and the type of strategy that best responds to this.

This created not only new opportunities, but also new risks. As this process continues, local authorities are having to think about their changing roles, so that they can maintain some status and function in private sector housing renewal. Local authorities need to be increasingly flexible in their responses to local housing conditions. Their role is seen as increasingly enabling – to encourage, support and assist private sector renewal activity, planning and land use processes. Part of this process is as direct actors, through individual grants, group repair, renewal areas, regeneration, clearance and enforcement activity. Part is as adviser, possibly in a partnership capacity.

There are several overarching government objectives to be met in developing and implementing housing strategies as described in Department of the Environment Circular 17/96 (DoE 1996) as follows:

- House renovation is primarily the owner's responsibility.
- There is an increased emphasis on involving the community.
- Local authority activity must demonstrate value for money.
- New initiatives need to help prevent decline.
- Renewal activity should be sustainable.
- Housing associations are seen to have an enhanced role in rehabilitation and development.
- Local authorities should make full use of their enforcement powers.

Changing role of housing strategy and options for private sector housing regeneration

Local authority housing strategies were traditionally about bidding for borrowing approval for their own house building programmes. They are now concerned with responding to the housing needs of the whole community, regardless of tenure or agency. Strategies are increasingly about forward thinking, the social and economic circumstances of current and future generations, looking at housing holistically and linked to need for employment, education and training, good infrastructure and regeneration. Current thinking (e.g. Goss and Blackaby 1998) suggests that this can best be achieved through a partnership of organisations by:

- listening to diverse views, choices and needs of local people;
- planning intervention in partnership with others; and
- accounting directly to local people for the success of the action taken.

A lot is expected of housing. It is now part of a wider contribution to ambitious social goals such as social exclusion, community regeneration and renewal, welfare into work, addressing the needs of vulnerable people, creating partnerships that work, community accountability and forging new relationships between local government and local people. A lot is also expected of local authorities' organisational structures and practices as strategies are increasingly implemented, in whole or in part, by external organisations. This brings new issues of management and quality controls of partnerships and commissioned services.

In tackling housing and its local communities, there is a sense that new things need to be tried, some of which may not yet even have been considered, but certainly a major thrust needs to come from the communities themselves. What do local people want and need? What might work? What needs to be addressed? In developing strategies, local authorities need to be mindful of, and sensitive to:

- regional differences;
- inequalities;
- supply and demand;
- regeneration;
- social exclusion;
- sustainability;
- urban and rural issues; and
- partnerships.

Private sector housing renewal is a rapidly changing policy area. Local authorities very basic, interventionist roles are being questioned as new partnerships are developing. There is less capital expenditure, with a break of the link of unfitness to mandatory grants, so more is required of house-owners through funding and behaviour changes. New possibilities for housing renewal need to be developed. This might include, for example, advice on home maintenance, tool libraries and so on. The general shift in private sector housing strategies has been toward encouraging greater use of private sector potential.

Central to this is constant review and revision of where available resources should be targeted and why. Local authorities have had to review fundamentally what they are doing and ask whether housing is the key focus, whether wider corporate initiatives have priority or if housing can be a key part of this, such as in area regeneration schemes. In doing so, renewal strategies need to:

- have regard to overall housing need and authority's housing policies and the role of other providers;
- set the strategy within the context of wider economic, planning and regeneration policies;
- cover problems facing all private housing; and
- define properties and identify most appropriate means of tackling the problem.

The success with which private sector housing strategies can be delivered is based fundamentally on available grants assistance and enforcement powers. Local authorities have the following issues to consider in delivering their housing strategies:

- Scope, application and potential of the main grants.
- Application of fitness standard and the role of enforcement.
- Promoting good maintenance.

- Works in default.
- Partnership(s) with home improvement or other agencies.
- Reactive responses to individual dwellings.

Funding renewal in the private sector

There has been a general shift of emphasis from local authority capital expenditure to other sectors, such as:

- voluntary organisations helping support home repair;
- single regeneration budget;
- funding for energy efficiency via agencies;
- low-, or no, cost surveys, such as by students carrying out HMO research;
- conversion of large properties into flats, with the owner as resident landlord;
- technological developments; and
- use of existing funding such as HEES, fuel utilities.

From duty to discretion

Local authorities have had to respond to continued changes in legislation, including a shift from mandatory to discretionary grants, with subsequent strategy development. Governments continue to state that owners themselves should have the major role in renovating their homes, and local authorities strategies have had to reflect this value, and look to alternatives for housing renewal. The current Housing Green Paper (DETR 2000) is proposing further change. Decreased capital funding and uncertainties threaten housing renewal activity for the future, as well as the actual percentage of subsidy being reduced. It would be difficult for a local authority, for example, to declare a 10-year renewal area without funding guarantees. Availability of discretionary grant assistance needs to be carefully targeted to the worst housing stock, or most needy clients, where it is likely to have the greatest impact. Criteria for grant eligibility needs to be clear.

Strategies need to take on board these issues and question the very nature of grants in being able to stimulate private sector investment. It is clear that an owner's lack of resources is not the only thing stopping them from carrying out repairs or improvements to their homes. New questions need to be asked about levering in private sector finance, and how successful, or not, this could be. What real incentives are there for the private sector (e.g. a landlord assurance scheme)? What measures can the local authority develop to stimulate borrowing and target incentives (including group repair, renewal areas, insurance schemes, saving schemes and so on)? What about on-going maintenance schemes? Are grants simply a one-off money hand out? Are they more than a snapshot of housing conditions? It seems timely for local authorities to take stock of the effect that their substantial housing capital investment is really having on the local housing stock.

Housing strategy and best value

Best value is the Labour Government's response to the now defunct compulsory competitive tendering process, where the drive to improve performance is seen as socially as well as economically efficient. It is a drive toward cost-effectiveness in local services; meeting needs and priorities of the local community; developing a culture of continuous improvement; developing staff in the necessary skills to change attitudes and operating styles and ensuring adequate resources and leadership; setting and meeting local targets and performance indicators and encouraging a long-term commitment to positive change, through the five 'C's:

- Compare performance.
- Consult with community.
- Compete with others.
- Challenge the purpose.
- Collaboration with others.

For further information, see DETR (1998), which looks at modernising local services through best value initiatives.

Best value should help develop community trust and give a voice to minorities by removing language, cultural and poverty barriers. It will provide more emphasis on local housing standards. Additionally, it should help empower tenants and owner-occupiers in terms of repairs, vandalism, crime and disorder, and community development. Part of the change could involve closer working with private sector landlords, a process that has already began in many areas to unleash the potential of the private rented sector.

Best value principles need to be incorporated into strategy documents and practice. Of course, many are already in place due to overlaps in policy areas resulting from wider changes in recent years. Moves toward greater local authority-wide strategies and collation of existing strategies through consultation and support have already highlighted the need for a review of resource use, improving services and 'advertising' the local authority housing function. However, there has also been a reduction in expenditure, as some local authorities have been wary of too much advertising, which may result in a demand they are unable to meet. In terms of private sector housing grants, it is timely that local authorities should fundamentally review what they are doing and the 'value for money' that arises from this expenditure. Consideration of such issues should help inform, evaluate and plan future actions. There is a need to look to other feasible solutions.

Dealing with the private housing sector is about intervening in the housing market. This intervention may be direct, by repairing existing stock through enforcement powers, financial support to owners through grant-aid and housing benefit. It may be more indirectly through consultation, participation and empowerment, although there are clear disadvantages in this approach alone in dealing with the private rented sector and enforcement normally proves necessary. It is difficult to see, for example, how area-based renewal might result from information and advice centres, leaflets and promotional events encouraging home maintenance issues, without adequate resource back up. There seems no real viable substitute for sustained financial investment.

Summary

- Local housing strategies were previously mainly about bidding for borrowing approval for local authority house building.
- They are now increasingly concerned with responding to housing needs of the whole community, regardless of tenure, agency or local authority.
- They are increasingly about forward thinking, social and economic circumstances of current and future generations, looking at housing holistically, linked to the need for employment, education and training, good infrastructure and regeneration – achieved through a partnership of organisations.
- Requirements under best value may prompt some changes in private sector housing strategies.

5.2 Key features in developing a strategy

Outline

An understanding of how to set about developing a strategy is key to ensuring its success in implementation. This section briefly overviews the theory of developing a strategy, whether it is to deal with individual dwellings, types of dwelling or a large-scale renewal scheme. The focus needs to be clear, include statutory and non-statutory action, priorities to be addressed, as well as being both proactive and reactive. Examples might be a strategy to develop special needs housing, the empty private sector, promoting energy efficiency, using sustainable materials, anticrime measures and specific issues such as dealing with listed buildings, radon and so on. The scope is endless, but some key points apply to each.

Developing a strategy

Developing housing strategies need not be too daunting provided that the remit is clear. There are four simple key questions that need to be kept in mind throughout so that the strategy achieves what it sets out to achieve. These are:

- Where are we now?
- Where do we want to go?
- How are we going to get there?
- How are we going to review our success?

There then follows several key features that need to be incorporated into the development of any new strategy. These need to be thoroughly considered to ensure that strategies are responsive to local housing needs through consultation, monitoring and review, reviewing existing local priorities and implementation plans of current and future strategies. Key considerations in developing a housing strategy are illustrated in Table 5.1.

Presentation of a housing strategy

A strategy document should be in plain English, setting out the local picture, past, present and future, and the strategic objectives and the strategy to achieve them. It should include a performance or operational plan and performance reports from participating agencies to check and ensure progress. It should also incorporate back-up information, containing a summary of the data and analysis behind the choices made (Goss and Blackaby 1998).

The strategic process should be well designed and presented, so that it can be readily explained in only a few sentences to anyone. It should indicate why decisions were made, and what it hopes to achieve and why. It should clearly illustrate how performance, or success, would be measured.

Summary

- There is increasing emphasis on achieving wider social goals through strategies.
- There has been a shift away from seeking government funding to

Table 5.1 Key features in developing a strategy

Heading	*Content*

Corporate approach	• Involve others both within and outside of the local authority
	• Process:
	  • establish a group;
	  • establish a brief;
	  • service delivery;
	  • active role; and
	  • obtain relevant approval (e.g. corporate and committee)

Set clear objectives	• What is required from a strategy – what does it seek to achieve?:
	  • unambiguous, clear, precise;
	  • understandable; and
	  • set targets – may be different for each objective

Assessment of need	• Comprehensive and rigorous • Use existing data • Sample the house condition survey – samples and outcome • Other sources?
Leadership and staffing	• Effective leadership throughout • Adequate staffing resources • Good knowledge and training • Team working skills
Evaluation of options	• Essential to be objective • Examples of methods • Number of properties and people • Costs • Effect on performance elsewhere • Characteristics of people

Realistic targets	• Targets must be realistic and achievable:
	  • timescales and outputs;
	  • credibility with the DETR;
	  • respect of residents, landlords, members, etc.;
	  • future funding; and
	  • motivation of staff

Monitoring and review	• A clear implementation plan should include systems to:
	  • collate and review information;
	  • compare performance against targets; and
	  • review mechanisms

Consultation	• Extensive • Mechanisms to reach as many as possible • Explain strategy and invite feedback • Respond to feedback

Best value	• The four 'C's:
	  • compare performance;
	  • consult with community;
	  • compete with others;
	  • challenge the purpose

attempting to deploy public and private sector resources to common social goals.

- Much time is needed to design an effective strategic process, including consultation, partnerships, collating information, testing options, and so on.
- It is important to consider how the strategy will translate into action and how this will be evaluated, as well as what remedial action is taken if it proves ineffective.

5.3 Developing a strategy for houses in multiple occupation

Outline

Houses in multiple occupation (HMOs) comprise the worst living conditions, frequently occupied by the most vulnerable members of society. As access into social housing has declined among some groups, access into low-cost accommodation, frequently substandard, has increased. Particular issues with HMOs were discussed in Chapter 4, illustrating problems of fitness and disrepair, amenities and inadequate means of escape from fire. This section reviews, in the light of criticism from the Audit Commission (1991), how local authorities might develop improved strategies to tackle conditions in HMOs. Whilst many local authorities, with support from their councillors, have good strategies in place, many still have a long way to go. This is particularly because there is continued and increasing market for such accommodation, where housing benefit is paid with little regard to condition, which is being exploited by some of the most unscrupulous landlords (see Section 2.3).

Tackling HMOs through strategic intervention

The National Consumer Council (1991) and Audit Commission (1991) reported that HMOs comprise the worst housing conditions and should, therefore, be a local authority priority. A decade on, the English House Condition Survey continues to show that conditions remain poor to some of the most vulnerable members of society, particularly in respect of fire safety, amenity provision, fitness and overcrowding. The private rented sector comprises the poorest housing stock, yet this sector is increasingly being seen as being able to contribute to meeting housing need. Government is encouraging greater use of this sector and encouraging local authorities to adopt both a strategic and enabling role. Local authorities are being encouraged to take a new approach to the private landlord and tenant, to increasingly involve them in the service offered in attempt to release supply in this sector. HMOs are seen to offer low-cost accommodation to people ineligible for social housing.

A major problem in tackling HMOs is that the exact scale of the problem is unknown, although estimates suggest that some 180,000 HMOs, more than 50 per cent of total HMOs, require attention (Audit Commission 1991). At current rates of activity, it would take 15–20 years to bring about improvements, not allowing for others falling into disrepair meanwhile. This points to the need for a proactive, efficient strategy that clearly defines priorities, based on the risk as well as the nature and scale of problem, sets clear objectives, sets targets, monitors and reviews performance.

There are many reasons as to why conditions in HMOs remain generally poor. Perhaps the main issues are the insufficient enforcement numbers to tackle conditions, combined with practical difficulties with addressing some very difficult enforcement issues. Local authority activity overall remains low, it is difficult to estimate accurately HMO numbers and there are variations in local authority efficiency and consistency. Tackling conditions in HMOs requires ample resource commitment by both central and local government, with support from elected members and management. Many local authorities have specialist teams to tackle HMOs, including EHOs and housing advice officers. Smaller local authorities not able to justify specific HMO teams are nevertheless taking an increasingly proactive stance to address some of the nation's poorest housing conditions.

There is much legislation available to local authorities in dealing with HMOs, including:

- Housing Act 1985 (as amended), covering fitness, amenities, overcrowding and over-occupation, means of escape in case of fire, control orders, etc.
- Housing (Management of Houses in Multiple Occupation) Regulations 1990.
- Environmental Protection Act 1990.
- Housing Act 1988.
- Housing Grants, Construction and Regeneration Act 1996.
- Housing Act 1996, covering registration schemes, administrative procedures.
- Other miscellaneous legislation.

There is currently a range of government guidance on HMO strategies and what is expected. These are listed in the References and Further reading lists and should be referred to when developing and implementing strategies.

The current green paper (DETR 2000) is proposing some substantial changes to legislation governing HMOs, but at the time of writing it remains too early to discuss these in detail.

Problems in dealing with HMOs

There are many reasons why it is difficult to address problems with HMOs. A major reason is their identification. Whilst frequently concentrated in city areas, they are increasingly springing up elsewhere in 'non-traditional' locations. Examples include hotels to accommodate the homeless, rural areas where housing prices have risen, and so on. This very fluid status, combined with their sudden growth and often 'hidden' status means that local authorities struggle to allocate resources. Many tenants are wary of reporting conditions, or indeed allowing local authority officers entry, for fear of a rise in rent, or eviction, or both. Many local authorities, perhaps more fundamentally, have inadequate resources, while others simply lack the political commitment and initiative to deal with them.

HMOs – establishing realistic strategies

In developing strategies, local authorities need to work with the range of information they already have on identifying and developing local risk assessment criteria as a realistic basis, which they have resources to tackle. There is no point in setting unrealistic targets, and managers need to bear in mind the fact that the numbers of HMO are very fluid. There are currently a number of issues in HMOs that now need thinking through for inclusion into likely future strategies. These include compulsory licensing (DETR 2000) as well as the ENTEC report, which proposes to link fire safety measures more closely to property risk (DETR and ENTEC 1997).

The priority in establishing a workable HMO strategy is to have full commitment from councillors, heads of department and operational staff and the organisational structure needs to be in place to ensure that the strategy is operable. This might involve key staff meeting, discussing and preparing an initial committee report detailing proposals and how these will be carried out, monitored and reviewed, whilst fitting alongside other strategies. At this point it is useful to indicate the overlap with other strategies, such as an anti-poverty or strategy tackling social exclusion. It is also useful to review how neighbouring authorities are tackling issues in HMOs, and by offering and accepting advice to and from others in a process of mutual learning.

The next stage is to begin to identify HMOs and occupants from as many sources as possible. This will help put an initial database in place to move forwards. Information, subject to data protection, can be available from housing benefit records, council tax, planning officers and local environmental health and housing case files, the electoral register, as well as local knowledge of HMOs. At this stage, it is useful to carry out an initial risk assessment and categorise HMOs, whether it currently complies with legislation, and so on, to have an idea of how long each might take to bring to standard. It is also useful to develop links with other agencies, such as homeless charities and landlords' forums for input.

The growth in numbers of HMOs means that risk assessment techniques have become increasingly important in prioritising resources to tackle conditions. Risk assessments tend to be local in nature to suit individual strategies, and are generally based upon size, means of escape in case of fire, state of repair, overcrowding and/or amenities. Officers involved should be fully briefed on how to assess risk assessment criteria, and what the outcome of their assessment would be in terms of priority ranking and re-inspection protocol in terms of the inspection rating total. High-risk HMOs should be tackled first in a planned programme of action. This programme should be realistic and flexible enough to allow for reactive visits and other staff commitments, as well as changes to the HMO profile over time. The programme planning should be regularly reviewed and appropriate amendments made as necessary. An example of a risk assessment sheet is shown in Figure 5.3.

The strategy statement is then necessary, whereby realistic objectives and activity are clearly laid out, alongside facilities for performance measurement and review. Each local authority will obviously adopt its own method of measuring performance, but it is useful to develop both quantitative and qualitative measures. This might include numbers of notices served and complied with alongside feedback from tenants to help improve services in the future. It might also involve a regular report to members on progress being made with the strategy, which is also a useful means of maintaining staff morale in the process. Visual aids such as before and after photographs and videos of conditions identified are useful to keep elected members informed and supportive.

A database should be established that relates to other local authority databases, so that it can be kept updated across departments. Key personnel should be identified to update the database. Consideration should be given to how and what data are to be collected and updated, so that DETR returns, for example, can be obtained as quickly as a mail-merged letter drop to all HMO tenants or landlords. The database should be a balance of being as simple as possible, but containing adequate information, so that it is not a chore to update, and staff can relate inputs to outputs. It should also be adaptable so that it can respond to changes in legislation. The database requires good management. Such a database can provide a basis for publicity and information flow to landlords and tenants, such as forthcoming changes to legislation (DETR 2000).

GENERAL		Tick	Score
Type of HMO:	Self-contained flats		20
	House let as bedsits		20
	Rented house		20
	Flat in multiple occupation		20
	Bed and breakfast		25
	Homeless persons' hostel		25
Vulnerable group:	Y		25
COMPLIANCE			
Repair:	Satisfactory		0
	Materially affects tenant's comfort		5
	Substantial disrepair		10
	Unfit		20
	Urgent repairs required		25
Level of occupation:	Satisfactory Y		0
	N		20
Provision of amenities:	Satisfactory Y		0
	N		20
Means of escape from fire	Satisfactory Y		0
	N		20
General management	Satisfactory Y		0
	N		20
MANAGEMENT			
	Highly confident		0
	Moderately confident		5
	Some confidence		10
	Little confidence		15
	No confidence		20
FURTHER ACTION			
	Notice		
	Prosecution		
	None at present		
INSPECTION RATING TOTAL			

Figure 5.3 Risk assessment pro-forma for HMOs

As with other strategies, corporate working is the key to success. Many local authority departments and other organisations have a role to play in HMOs, including housing advice, homelessness, environmental health, building control and planning. Working together will help mutual learning of each others' roles, as well as encouraging discussion on new initiatives, such as a scheme for housing the single homeless. Corporate working will help reduce bureaucracy and an information leaflet can be prepared to advise tenants on where to go to get help for specific issues. Referrals may come from outside bodies, or the local authority may refer a tenant to an outside organisation, which may be better placed to meet a specific need.

Working with landlords and tenants is fundamental to the success of an HMO strategy. Whilst in some HMOs, there is no option but to enforce legislation, there is some scope elsewhere to work with landlords to improve standards. Some may, for example, be willing to carry out additional works if they receive a grant for means of escape works, or to upgrade energy efficiency, and every little improvement adds to housing conditions overall. If service improvements can be put in place that benefit landlords and tenants alike, things are likely to move forward in a positive manner. However, the fact remains that this is the poorest sector of housing stock, and strategies that rely solely on education and persuasion are unlikely to address the conditions found unless backed up by a sound enforcement protocol based on effective works in default and prosecution procedures. However, if some form of regulation can be put in place by education and persuasion, it will free-up local authority resources to tackle higher risk HMOs.

The enforcement aspect of HMOs is key to tackle the worst of the nation's housing stock in an effective and efficient manner. Staff need to be trained in all aspects of residential legislation governing HMO conditions, as well as an understanding of landlord and tenant law and where to seek further advice. Officers dealing with HMOs need a sound knowledge of fitness, disrepair and nuisance. They also require an administrative and technical understanding of, and ability to apply, specific HMO requirements such as means of escape and amenity provision. The strategy needs to clearly state what the enforcement protocol will be, and what will happen, and when, if it is not met. Again there needs to be some flexibility built in to allow for the issues that invariably arise in HMOs, such as it falling into and out of occupation, or difficulties in gaining access and so on.

A summary of the keys to a successful HMO strategy outlined above is contained in Table 5.2. An outline of how a small local authority established a proactive and effective HMO strategy is outlined in Box 5.1.

Box 5.1 Theory into practice – starting from scratch with HMOs

A small district authority had never paid a great deal of attention to its HMOs, and there had been little political interest in this area of housing. Following on from the Audit Commission Report, Department of the Environment surveys and similar, and continued proactive efforts by the environmental health department, the situation gradually began to change.

Key EHOs and other related housing officers began to meet regularly interdepartmentally to exchange knowledge, and between them identified an initial number of possible HMOs. This process also served to help various officers from planning, housing and environmental health to understand each others' roles and functions, as well as overlaps and possible contradictions in some of their work. It enabled them to put initial paperwork and possible procedural set-ups in place to get the ball rolling. Some initial joint visits took place to help in mutual learning and to develop good working relationships.

Possible HMOs were quantified in various ways, but the local authority was surprised to find the volume of information it already held. Much work was done to identify the location and extent of HMOs, including manually going through printouts of housing benefit and council tax records, subject to data protection – a time-consuming but fruitful task. Once this initiative got off the ground, several officers mentioned that they had always had suspicions about particular properties possibly being HMOs, but previously had never the time to visit. Certain streets were targeted for external surveys, looking for things like extra dustbins, doorbells, general appearance of property, different net curtains at each window, etc. The new HMO strategy initiative had the added advantage of starting positively to change many officers' attitudes to, and interest in, HMOs. Key officers began to develop standard forms, sheets and letters that would be required for inspections and revisits. An initial number of HMOs were established and the next course of action discussed with the group.

A registration scheme was proposed to add impetus to the initiative and was taken to the housing committee for authorisation. This also proved a useful opportunity to inform some councillors of some issues in HMOs. Generally, the proposal received much support, although some councillors were wary of 'too much bureaucracy in the private rented sector'. The scheme was formalised and the series of standard letters and other documents that had been prepared was sent out to the identified possible HMOs in manageable batches, as well as publicity on the scheme to local estate agents, housing charities and others who may have an interest and further information.

A surprisingly high response was received from the initial letter, and details were collated and non-returns followed up. This initial action started the programme of risk assessing HMOs and enabling newer members of the technical and administrative staff to be introduced to HMO law and practice. The HMOs were risk-assessed and data inputted to meet the revisit programme recommended by the Audit Commission.

This enabled an overview of HMOs in the district and a realistic, planned programme of inspections to be developed based on numbers and staff availability. The private sector housing strategy was updated accordingly. The emerging strategy also provided the basis for closer working with external organisations, including charitable homeless groups and the fire authority, which proved very beneficial.

Information was discussed with the local CIEH housing group. There were EHOs and technical officers here who had vast HMO experience which they were willing to share with the group. This served to generate some interest, and promoted further sharing of information, general inspection sheets, leaflets to tenants and landlords' standard phrases, procedures, and so on, and the increased numbers of officers allowed tailored training at a more economic cost to all.

The database was designed in terms of what outputs were required for HIP statements, DETR returns and the local authority's performance indicators, landlord and tenant details, and so on, rather than what could be input into it. The process of identifying and risk-assessing added to the workable database, which could be readily adapted to include forthcoming initiatives, such as proposed risk criteria changes following the ENTEC report, and proposals for HMO licensing.

Table 5.2 Key issues in establishing a strategy for houses in multiple occupation

- Organisation and commitment by councillors, heads of department and operational staff
- Identification of HMOs and occupants
- Strategy statement, performance measures and review
- Establishment and management of HMO database
- Priority planning, risk assessment and programme of action
- Corporate working within authorities and outside bodies
- Working with landlords to help secure upgrading
- Services to tenants, e.g. advisory services and rent deposit schemes
- Publicity and information flow to both landlords and tenants
- Full use of appropriate powers, including the purchasing power of housing benefit and an explicit statement of what enforcement powers will be used in what circumstances.

Source: based on DoE (1995)

Summary

- HMOs comprise some of the worst living conditions and local authorities need to develop and implement effective well-thought-through local strategies.
- Successful strategies need to take many issues on board, including wide involvement, support and management, and they require a regular review.
- Proposed changes to legislation need to be thought through and incorporated into existing HMO strategies at the earliest oppor-tunity.

5.4 Developing a strategy for vacant dwellings and optimising the private rented sector

Outline

There are more vacant properties in this country than households, so in theory no one need be homeless. However, the location of that vacant accommodation is frequently in the wrong place for people to live in. There are issues about why properties are vacant, why people are unable to access accommodation, which need to be acknowledged in developing a strategy to tackle vacant properties on a local level. The government is keen to encourage this accommodation to be brought back into use and for local authorities to develop suitable strategies to do so. This section looks at some of the possibilities of bringing vacant properties back into use, how it might be done, and how it can overlap into other strategic areas, including environmental and housing need.

Most vacant properties are in the private sector and offer enormous housing potential, which local authorities may be able to encourage by targeting grants here, and attaching conditions to gain nomination rights. This section looks at what can be done, and how, to tackle not only just homelessness and housing need, but also to turn around the blight caused by the effect a vacant property can have on a neighbourhood.

This section looks at some issues surrounding empty homes, before turning to the local authority role in bringing them back into use and other ways of encouraging the private rented sector to provide accommodation locally.

Vacant properties – a national disgrace?

English empty dwellings increased by 42 per cent from 539,000 in 1983 to 764,000 in 1993. This represented around 4 per cent of total housing stock, mainly being in the private sector. There are many reasons for this rise in the private sector, including complex reasons of ownership, changes in the housing market, lack of resources or interest and so on.

There are approximately ten times the number of private sector than local authority empty homes (JRF 1994). Some of these are transitional, whilst others are problematic vacants. Some examples of vacant properties are shown in Figures 5.4–6.

More recent figures from the DETR HIP 1 Returns submitted by local authorities (cited in EHA 2000) show some progress in tackling empty properties in the private sector since 1996, but an increase in vacant local authority and housing association homes. There were 667,000 vacant private sector homes in 1996, and this was reduced to 636,000 in 1999, representing 3.63 per cent of total housing stock. This figure is still very high and there remains a long way to go. It is only viable action at local level that can impact these national figures. Government figures showed that, in 1998, 753,000 homes stood empty whilst 100,000 households were accepted as homeless. Whilst not all of these vacant properties would be suitable in terms of location or design to meet this need, the numbers are still unacceptable.

Figure 5.4 Vacant property. This illustrates the external areas of a flat that had been vacant for around 8 years. Despite being a waste of a potential three-bedroom family home in an area of demand, it had attracted vandalism, arson and a pigeon population, causing nuisance to neighbouring properties. The flat was brought back into use when the local authority instigated a vacant property strategy. The local authority initially offered advice and assistance, but this did not have the desired effect. With the benefit of a vacant property strategy in place, the local authority was in a position to instigate compulsory purchase with subsequent management by a housing association. The threat of this encouraged the landlord to apply for grant assistance to renovate the property on the understanding that it would be let with local authority-nominated tenants for 5 years

Figure 5.5 Vacant property. This illustrates the kitchen area of the flat shown in Figure 5.4. The flat did not require extensive internal works, already having all standard amenities, although it required some attention. Necessary works were relatively inexpensive in relation to the cost of having a property left vacant whilst having to rehouse homeless families in temporary accommodation

The Empty Homes Agency (EHA), an independent housing charity, was established in 1992. Its main objectives are to draw attention to the scandal of empty properties, devise solutions and disseminate good practice on how to tackle them. It focuses on local solutions to tackle

Figure 5.6 Vacant property. This illustrates a listed building that had been vacant for more than 10 years. The owner seemed to have lost interest because of the likely high cost of renovations that needed to be done to meet listed planning requirements. The local authority approached the owner informally at first, but the situation did not change. It then made a more formal approach, the carrot of possible grant-aid tied to nomination rights, and the stick of possible compulsory purchase should the property remain vacant. The owner did not want the responsiblity of retaining the property, and responded by putting the property on the market and a new owner soon purchased it. The house was fully renovated to a high standard as a terrace of three properties to be let on assured shorthold tenancies

what amounts to a national disgrace. EHA (2000) argues that empty homes are bad for communities for the following reasons:

- Deny homes to those in housing need.
- Create ghost towns in inner urban areas.
- Increase the pressure to build on green-field sites.
- Attract crime and antisocial behaviour.
- Blight neighbourhoods.
- They cost everyone, through loss of council tax revenue and devaluation of surrounding properties.
- Wasted asset for owners, wasted resources for the homeless and those otherwise in unsatisfactory housing.

Bringing vacant properties back into use also has a very visual real impact on the local environment, and bringing several back into use can start the process of developing confidence and regenerating an area.

Using the vacants to meet the need

There is usually a reason as to why a property is vacant, particularly if it is a long-term vacant. Most vacant properties are in the private sector, so there is enormous potential in this sector to start bringing such properties back into use.

The private vacant sector has several advantages. It already exists and a lot of vacant property can be brought back into use at relatively low cost in comparison with whole-scale redevelopment, particularly if there is a local demand for accommodation. It offers some variety and flexibility in accommodation type and location. There are many local authority structures and personnel already in place who can assist with the private sector, including housing advice, environmental health and landlord forums. They can offer flexibility of initiatives to encourage use and funding possibilities, such as grants and housing benefit, which the private sector does not cater for. Prioritising available resources into tackling vacant properties can have a substantial impact on the problem, particularly when grant activity is combined with nomination rights to access the private rented housing market.

Housing supply and demand

Housing supply and demand is largely determined by social, historical, political, economic and investment processes in local and national housing. Constraints can arise from 'interpretation' of legislation and government regulation, and also from organisational arrangements and culture of a housing organisation. Policy-making and implementation, subject to suitable housing availability, are key. The imbalance in housing availability and suitability is multifaceted. It results from an increase in population, a marked increase in the elderly population, changes to family structure, the availability of private sector housing including affordability, sales of local authority housing and differences in local authority interpretations of 'homelessness'.

Many local authorities are working to bring vacants back into use and their strategies to do so will invariably involve both a carrot and stick approach. There are several phases to bringing properties back into use, and local authorities have a range of persuasion, educative and enforcement powers at their disposal, which include the following:

- Housing Act 1985 (as amended) – duty to survey annually stock and to determine housing strategy; duty to act in respect of unfitness, subject to MSCA; power of compulsory purchase.
- Housing Grants, Construction and Regeneration Act 1996 – potential for discretionary grant-aid targeted toward vacant properties, with a nomination of conditions attached.
- Building Act 1984 section 79 – to deal with ruinous and dilapidated buildings and neglected sites detrimental to the amenities of the neighbourhood, with powers to repair, restore or demolish buildings and to remove rubbish.
- Town and Country Planning Act 1990 – powers to deal with land use where it is adversely affecting the neighbourhood.
- DETR Planning Policy Guidance Note 3 – to encourage sustainable local authority land use plans, which make greater use of existing brown-field sites and reuse of existing accommodation in preference to green-field developments (DETR 2000).

The strategy may be a way of dealing with a vacant property for its own sake – to prevent further decline both of the property and in the wider neighbourhood; or it may be more directly to allow the local authority to gain nomination rights to the dwelling by offering a grant and applying such a condition for 5 years – this may, for more reasons than one, present considerable value for money when the alternative cost of accommodation is considered for a homeless household. It is difficult to anticipate what effect a vacant strategy might have in respect of a single property – the local authority intervention may result in the property being sold to another owner-occupier, but at least it is brought back into circulation, with the added advantage of council tax being paid.

Proposals for new legislation

Some do not feel that current local authority powers go far enough and that new legislation is required if the government is serious about bringing properties back into use. David Kidney MP introduced a Private Members' Bill to tackle empty houses in July 1999 (EHA 1999). The Bill proposed that dealing with empty houses was an essential part of wider government commitment to tackle area blight and disrepair. The Bill proposed new duties and powers really to make an impact by, first, introducing a new duty for local authorities to establish and publish a comprehensive corporate strategy to reduce empty properties. This strategy would include housing need figures and numbers of empty homes, targets for reducing empty homes, actions to be taken and details for monitoring success. The Bill also proposed to extend compulsory purchase powers for empty properties and a discretionary power for local authorities to charge owners variable council tax for long-term vacants.

Renting privately – the local authority role

Local authorities do not have a duty to provide accommodation for everyone who is homeless (see Section 2.5). They have to make enquiries as to why the applicant is homeless, or threatened with homelessness, and determine an appropriate course of action for each case. Their response may range from providing advice and assistance, to securing permanent accommodation for applicants in cases of where there is priority need. In between these two extremes, there is some possibility to assist potential private sector tenants in securing accommodation in that sector, through encouraging vacant properties back into use, with or without a landlord assurance scheme.

There is clearly less local authority housing to meet growing need and local authorities are having to look further than traditional forms of social housing to meet growing demand. Despite these initiatives, there remains much criticism that resources are simply inadequate to meet growing demand and mass social housing provision is still required. Even with the release of capital receipts, this looks unlikely to happen in the near future. Local authorities are, therefore, tasked with making the best use they can of existing local accommodation and private sector landlords, through initiatives such as the landlord assurance scheme. Many have worked across departments to produce information packs, offering advice and assistance to those who may have properties to let.

The landlord assurance scheme is a system whereby the local authority 'approves' a landlord and the property offered for rent. It therefore serves as a means of regulating this sector to some extent, but is already likely to attract better landlords and properties. This scheme is commissioned and promoted by local authorities and it is voluntary in nature. Properties must meet all legal standards, and are generally inspected by environmental health departments prior to letting and are subject to management standards. Properties registered within an existing HMO scheme can be automatically accredited.

There are benefits both to owners, tenants and the local authority in introducing a landlord assurance scheme. For owners with suitable quality properties to let, there is the potential for fast-track housing benefit application processing; access to a rent guarantee scheme and a seal of approval through a recognised award. For tenants, there is safe and secure housing that they might not otherwise be able to access; proper tenancy agreements with tenancy rights protected and the possibility of rent guarantee assistance to secure accommodation in the first place. To the local authority, there is potential to encourage a positive image of private rented sector; potential to encourage enhanced standards; demonstration of a partnership approach, which is becoming increasingly important in HIP bids and improved opportunities for rehousing options locally.

Mechanisms for review will vary in accordance with the nature and objectives of the initial scheme, and feedback can take many forms. In trying to use the private sector to meet need, there has to be some form of incentive to the landlord, or they simply will go elsewhere. This may be the offer of housing benefit paid promptly, or someone else to manage the tenancy for them. Such information can be unleashed through landlords forums', tenants' meetings, feedback from surveys and so on. But perhaps the best way to measure success for such a strategy is to simply witness long-term vacant properties being brought back into use, and once again being occupied, which elected members – as well as the public – can see for themselves, with the awareness that their local authority made it happen.

Summary

- Vacant properties are a wasted resource when so many are without a home.
- They can cause local environmental decline.
- Local authorities need to establish sound strategies comprising education, persuasion and ultimately enforcement through compulsory purchase, to bring long-term vacants back into use.
- The vacant property strategy can operate alongside other strategies, so that, for example, grants can be targeted to this sector allied to the local authority nomination rights, and that strategies such as a landlord assurance scheme can also encourage initial or reuse of the private rented sector.
- Local authorities need to be innovative in their approaches to optimising private sector potential, and can gain information and ideas from many sources.

5.5 Urban regeneration

Outline

A history of private sector housing renewal was reviewed in Section 2.1 and it is this history that has flavoured current activity in this sector. What generally emerges is a separation of renewal across tenures, due largely to funding opportunities and ownership differences. Much social housing has had specific programmes for resource allocation, such as estates action, housing action trusts and, more recently, the estates renewal challenge fund.

Until the Local Government and Housing Act 1989, private sector housing renewal was based on small-scale housing schemes with no legislative scope for wider area regeneration. This Act introduced the concept of renewal areas to take on board wider regeneration issues, essential to sustainable renewal programmes. These were developed further through the Housing Grants, Construction and Regeneration Act 1996, with government guidance in the form of Department of the Environment Circular 17/96, *Private Sector Renewal: A Strategic Approach* (DoE 1996), which for the first time provided detailed guidance on area-based action, incorporating issues wider than just housing rehabilitation.

There is currently other potential resources in the form of single regeneration budget, which has to be competitively bid for. Current regeneration strategies have to take on board this process of bidding for a declining base of funds, and accessing resources elsewhere, such as the European Union as well as becoming increasingly innovative in new initiatives in urban regeneration.

This section looks at the process of urban decline, how this is understood, and subsequent policy development in terms of political ideology to intervene and reverse that decline. It looks at legislation and funding available to local authorities to deal with urban regeneration and how this is evolving.

Urban regeneration – the policies, legislation and resources

What is regeneration?

Regeneration is wider than just housing rehabilitation. In its urban context, regeneration is about the improvement of a distinct geographical area by tackling a wide range of factors causing disadvantage and decline. Regeneration activities range in scope from small-scale community development projects to large-scale regeneration with substantial capital investment (Hooton 1996). The nature and extent of regeneration is frequently determined by funding and other resource opportunities. Urban regeneration is largely governed by the extent to which the State chooses, or is able to involve itself in a market-based system that has ultimately failed to be sustainable. This State involvement is mainly subject to resource availability, particularly financial.

Whilst it would be encouraging to claim that regeneration schemes are always successful and sustainable, this is not always so. Even the government's own publication (DoE 1994) points to the facts that some schemes have not been successful in the longer term and their research illustrates the need for sustainable renewal programmes based on a sound and guaranteed financial footing, which allows for flexibility and partnership approaches, including clearly defined local authority and community roles combined with programme coherence across and within government departments. The report also suggests the possible development of an urban budget administered at regional level to improve coordination across programmes and departments. This is clearly in line with European Union thinking and resource allocation. The key objectives to include in effective urban regeneration policy are illustrated in Table 5.3.

What then emerges is a need for honest appraisal of the current urban situation before embarking on a regeneration programme and the following questions become key:

- How is the area of decay defined? By whom?
- What policy will be put in place, and is this actually likely to be effective?
- What is its goal? How will this fit in with prevailing market forces in the area?

Table 5.3 Key objectives to include in urban regeneration policy

- Economic revitalisation to promote industry and commerce as well as creation of employment in the building sector
- Decentralisation of power to local government and local community involvement in the renewal process
- Involvement of public–private partnerships
- Improvement in use of infrastructure
- Improved quality of life and environment, including social renewal initiatives and energy conservation
- Rehabilitation of historic buildings and districts to maintain character
- General residential improvements with a guarantee for residents to stay after improvements; maintenance of low-cost housing; transfer of rented dwellings to owner occupation or cooperatives; better special needs housing, particularly for the elderly and disabled

Source: based on Skifter Andersen (1999)

Process of urban decline

There is not scope here to analyse fully the economic process involved in urban decline, but to provide a brief overview of the changing nature of the urban arena. Balchin *et al.* (1995) argue that urban regeneration is a reaction to changing requirements and demands placed on large conurbations and the inevitable process of obsolescence and reconstruction. They add that redevelopment, rehabilitation and relocation of services are all part of the urban growth and renewal process. Such change essentially relocates urban activities, spontaneously regenerating some areas, but causing stagnation and decline in others, and inevitably affecting wider issues such as transport and infrastructure. Areas in decline may fall into further decline as market investment withdraws due to risk, causing local decay. The process tends to be a downward spiral, with a blighting effect on neighbouring areas, with depressed property values and little incentive for improvement.

The effect is similar for housing, as older dwellings are vacated by higher-income groups and re-occupied by lower income, or benefit-dependent groups experiencing high levels of unemployment, with a possible shift to multiple-occupation, serving as an inexpensive form of residential accommodation. Such rented accommodation, where heavily subsidised by housing benefit, can also serve to support low wages and temporary labour, and, of course, the poverty trap. The urban environment has therefore become increasingly temporary and there is little market incentive to invest in rehabilitation. Some see this as a key feature in intervening in private housing, which is already more difficult to address than social housing because of dispersed ownership and low property prices (Simpson 2000). The combination of low demand and decay inevitably effects land values.

Whatever their ideology, governments have involved themselves in some form of subsidised urban regeneration, and associated intervention in the market in the form of local taxation, for example, for around a century. Such intervention has taken many forms, but became particularly prevalent in the 1960s and 1970s in whole-scale clearance and redevelopment. Such intervention and municipalisation is no longer seen as politically desirable or indeed sustainable. The emphasis is now on partnerships approaches, with communities themselves taking a key role, and mixed funding opportunities – a mixture of the market and the State working together in urban regeneration schemes. Regeneration also requires some process of getting people involved – whether by persuasion, assistance (such as through grant-aid) or compulsion (such as through enforcement notices), and it is up to each local authority to determine a suitable way forward.

Cities are a primary source of economic progress and wealth creation, but at the same time suffer many social costs brought about by urban change including industrial decline, poor housing, unemployment, crime, social exclusion and so on (Oatley 1998a). Cities therefore contain both opportunities, and issues arising from a combination of economic, social, political and ideological changes over time. There is pressure to retain a competitive edge in a global economy, and a simultaneous need to address urban deprivation. Recent years have seen major economic shifts toward high unemployment, short-term contract working, etc., which has been in part responsible for an increase in inequality, welfare dependency and social exclusion. Political responses have included increased partnership working, centralisation of government and new management practices, brought about largely by New Right ideology, a rolling back of the Welfare State and principles of competition and privatisation (Oately 1998a). Both the nature and history of the urban environment influence urban regeneration policies, but political ideology also plays a key role.

Land prices, and consequently land use, play an important role in any urban regeneration programme, and invariably has some effect on the commercial–residential pressures in an area. The nature of local government, and the financial, political and community power it is allowed to have in relation to central government, can help influence planning and land use patterns and how much of a role housing can play in the regeneration process, rather than being a peripheral issue in urban regeneration. Skifter Andersen (1999) points to the following justification for the State to intervene in improving poor housing conditions:

- Provision of decent housing conditions for all.
- Long-term preservation of housing stock.
- Prevention of health problems caused by substandard housing.
- Provision of savings to social care services.
- Making a contribution to broader regeneration programmes.

Clearly, these all have different implications in terms of strategies and resources required to achieve them and the role that the State takes in meeting required objectives.

Urban regeneration – the local authority role

In terms of regeneration, the government has developed and administered complex systems directed to meeting specific objectives across different housing tenures. As far as private sector housing is concerned, the first real area-based schemes were introduced in the form of General Improvement Areas and Housing Action Areas by the Housing Acts 1969 and 1974 respectively. Although theses schemes had some success, there were criticisms that resources were not targeted to the worst properties and the most in need, but that some landlords were exploiting the grant system and that the grants caused gentrification in some areas. Enveloping schemes followed, whereby grants were targeted to whole blocks of houses for external renovation which was seen as more cost-effective, and had the objective of improving confidence in an area and encouraging owners to invest in further works. Such schemes were housing-based, and largely overseen by environmental health officers within a local authority. Such schemes had little scope to bring wider issues, such as poverty, deprivation, exclusion, employment and capacity building opportunities into the equation.

It was not until the Local Government and Housing Act 1989 (where grants for fitness were means tested and mandatory), and more recently the Housing Grants, Construction and Regeneration Act 1996 (where such grants became discretionary), that legislation provided for renewal areas as a new means of pioneering urban regeneration. Renewal areas provide the administrative framework to deal with social, economic, environmental and housing problems in an area of mainly private sector housing, which involves a range of strategies to encourage sustainable renewal. There is no dedicated funding, so local authorities have to identify their own resources as part of establishing a renewal area. There are currently one hundred renewal areas in England (DETR 1999a). Such area-based approaches have many benefits over other types of strategy, including:

- tackling both poor housing and its social environment;
- developing long-term sustainable partnerships;
- the possibilities of bidding for other funding, such as the Single Regeneration Budget;
- the stimulation of private investment through developing confidence in the area; and
- providing a local strategic framework for group repair, renewal, clearance or targeted renovation.

Renewal Areas superseded GIAs and HAAs and provided a new framework for a comprehensive approach to improve housing as well as the local socio-economic environment through developing partnerships, extensive regeneration and mixed-use funding and to provide maximum impact by increasing community and market confidence to help reverse decline. A renewal area would contain from 300 to 5,000 dwellings, of which 75 per cent would be privately owned and 75 per cent of which would be in poor condition. In addition, 30 per cent of households would be in receipt of income-related benefit. Establishing a renewal area is based upon a comprehensive area appraisal prior to declaration, using Neighbourhood Renewal Assessment as detailed in Department of the Environment Circular 17/96 (DoE 1996). This provides the framework for 10-year resource identification, publicity and consultation required, before seeking Secretary of State approval. After declaration, local authorities need to ensure a flow of information, an implementation plan, mechanisms for monitoring and review, a strategy to suitably end the scheme and importantly give consideration as to coordination of action with other local authority schemes. This is summarised in Table 5.4.

For a flow chart on the procedure, see Bassett (1998: FC118).

Group repair and clearance can be used within renewal areas to help achieve required objectives. The aim of a group repair scheme is to renovate blocks or terraces with mixed funding, within or outside of a renewal area. The objectives are to improve the external appearance of dwellings and target resources on area basis. This is thought to encourage owners to then invest their own finance in carrying out further necessary repairs and on-going maintenance, although the extent to which this actually happens is unknown. Clearance areas (as outlined in Section 4.3) seek to replace worn out housing and can be an important dimension of housing strategies when considered alongside possible renovation as the most satisfactory course of action. The issues to be addressed through clearance are sensitive ones concerning the local community, since compulsory purchase orders can result in blighting the area.

Impact of regeneration

To be successful, regeneration schemes need to be able to tackle broader urban problems and the extent to which poor housing conditions are part of a general social and economic process in a geographically defined

Table 5.4 Establishing a renewal area

Before declaration	*After declaration*
• Comprehensive area appraisal • Use NRA process • Identify resources for 10-year programme • Carry out publicity and consultation • Prepare a report • Seek Secretary of State approval	• Provide adequate information and advice • Prepare an implementation plan • Monitor and review activity • Prepare an 'exit' strategy for the end of the programme • Coordinate renewal area action with other local authority schemes

Source: based on DoE (1996)

urban area. It can be difficult to find a way forward in some areas where traditional industries have disappeared, leading to an ageing population and lack of young inflow, causing long-term decline which can cause changes to local housing markets due to low demand for owner occupation and an increase in private renting, with less funding for private sector renewal. There is a need for investment that some regeneration programmes are not meeting (Brooks 2000). Administrators need to decide the feasibility of rehabilitation (of existing stock) versus renewal (providing new housing stock, which is likely to have a greater impact on residents) and how social and economic issues can be realistically addressed (Skifter Andersen 1999).

The impact of housing renewal legislation from the 1989 Act has been mixed. One major difficulty in area-based schemes has been the means testing element, and encouraging wide take up to improve an area. One of the difficulties in declaring renewal areas has been the lack of long-term funding guarantees (Leather and Mackintosh 1993, Leather 1999, Skifter Andersen 1999). This has resulted from a shortage of capital resources, inadequate local authority funds to meet other forms of grants and the inability of the means testing system to take account of an applicant's out-goings and the lack of grants to the private rented sector, where conditions are poorest (Leather *et al.* 1994). The legislation has had some impact on preventing further decline in housing conditions, but has made little overall impact on stock improvement. There were also problems with grant distribution nationally, with some areas of housing stress receiving proportionally less than elsewhere (Balchin 1995).

Competition in funding opportunity

Urban policy has altered radically since the 1979 election of a government with a New Right philosophy and ideology. The focus of the New Right shifted away from local management toward contracting and competition in the urban regeneration arena, and this has been largely retained by the New Left. Part of this change has been a move away from traditional property-led regeneration, toward exclusion and economic competitiveness. The philosophy is manifested and characterised through funding opportunities, new forms of local authority management and partnership working (Oately 1998b).

The ideas of competition and challenge were introduced into private sector housing (and wider urban) regeneration in the 1980s and has become a key political tool in resource allocation. The City Challenge initiative was introduced in 1991 to fund 5-year programmes of comprehensive urban regeneration. To be successful in achieving funding, local authorities would have to illustrate how their proposals were comprehensive and ambitious, contained partnership proposals with the private sector, provided for local community participation and that they had arrangements in place for implementation and delivery (Balchin 1995). Through the bidding process, thirty-one winners each received £7.5 million over a 5-year period (Oatley 1998b). This amounted to additional potential funding toward urban regeneration strategies. Renewal initiatives since 1989 are summarised in Table 5.5.

City challenge was incorporated into, and superseded by, the single regeneration budget (SRB), along with other funding initiatives from other government departments, such as estate action, development corporations and housing action trusts, in 1994. The introduction of SRB was the most significant re-organisation of urban policy since the 1978 Inner Urban Areas Act and a proportion of SRB was made available for new regeneration schemes designated under the challenge fund, delivered by regional offices to be more comprehensive and accessible (Oately 1998c, DETR 1999b). The fund sought to recognise problems of poverty,

Table 5.5 Initiatives in private sector housing and urban regeneration

Date	*Initiative*	*Key regeneration purpose*
1989	Local Government and Housing Act 1989	Introduced mandatory means-tested grants based on fitness; introduced home repair assistance; introduced renewal areas and group repair schemes; redefined clearance areas and action for individual dwellings
1991	City Challenge	Programme to rehabilitate housing and commercial areas and at the same time to provide training for employment
1994	Single Regeneration Budget	Combination of some twenty previous urban aid budgets administered through regional offices to improve economic and industrial competitiveness, employment, social, and physical environment and quality of life. Bidding, competitive process. Initiatives to be comprehensive and part of a wider strategy incorporating partnerships, providing added value and value for money
1994	Lottery Funding	Added value-type funding to support urban regeneration initiatives
1995	Regional Challenge	Resource derived from EU Structural Fund for competitive bidding for public–private partnership projects, to stimulate innovative regional developments and maximise private sector contributions
1996	Housing Grants, Construction and Regeneration Act 1996	Discretionary grant-aid for renovation; further proposals for renewal areas and group repair schemes; home repair assistance extended
1997	Social Exclusion Unit	Established to coordinate policies to tackle social exclusion

Source: based on Oatley (1998)

isolation, community breakdown and industrial decline and associated issues by targeting some areas, based on need, not previously receiving priority for assistance.

SRB provides resources to support regeneration initiatives undertaken by local regeneration partnerships. SRB is very much about local regeneration and capacity building through relevant partnership support. It seeks to reduce inequalities and areas and groups, and encourages best practice as well as value for money. SRB allocation varies according to local circumstance, but needs to include some or all of the following as identified by the DETR (1999b):

- Improvement of education, skill and employment prospects locally.
- Addressing social exclusion and improving opportunities.
- Promotion of sustainable regeneration through improving and protecting environment and infrastructure, including housing, commerce and industry, and social problems such as crime, drugs and community safety.

SRB is administered through regional development agencies, except in London where the London development agency has this function. Bidding in rounds one to five led to the approval of 750 schemes amounting to £4.4 billion for up to 7 years. This is likely to attract over £8.6 billion of private sector investment, and make funding from the European Union more likely. Such partnerships are likely to create or retain 790,000 jobs, complete or rehabilitate 296,000 homes, support over 103,000 community organisations and 94,000 new business ventures (DETR 1999b).

Such competition in accessing funding has radically altered the way in which policies tackling urban decline and social disadvantage are formulated, funded and administered. Urban and rural locations alike can bid for the same budget – a shift away from an urban regeneration fund. Competitive bidding and funding has altered the way in which local governance, management, local representation and leadership have had to develop and operate, particularly in terms of new partnership (Oatley 1998a). The shift has been from local governance toward central control over funding allocation and altered previous methods of allocation and at the same time seen reduced government funding (Balchin 1995). There remain many arguments about the nature, extent and accountability of SRB funding.

The EU has had a growing influence on domestic policy, with some £10 billion allocated to the UK through the European structural fund between 1994 and 1999 (Oately 1998a). As government resources have declined, local authorities are having to look elsewhere for funding, and the EU offers some potential for urban regeneration projects. Funding has been available from the EU in the form of regional challenge – launched in 1995 – modelled largely on city challenge. This resource is allocated from the domestic allocation of the EU structural fund and must be competitively bid for by public–private partnerships. A total of £160 million was allocated in 1995, and repeated in 1997. This accounts for 12 per cent of the domestic fund allocation from the EU. The EU is seeking to develop the role of cities in the global economic context, and the current Labour Government is supportive of this policy.

The philosophy of competition and bidding for funds has to a large extent been retained by the current Labour Government, with a general trend in emphasis away from pure economic efficiency concerns toward cooperation and best use of resources (Oately 1998b). However, there has been a greater emphasis on issues of social exclusion, and interest shown in the social and economic consequences of long-term unemployment (Oatley 1999a), but delays in the government's urban White Paper and adequate strategies, funding and freedom in place to pioneer urban change (Hatchett 2000). The current approach is social democratic, and resources remain severely constrained. It is difficult to see how effective such an approach will be where a radical, if not revolutionary, approach is required to deal with if some of the root causes of urban decline, such as disadvantage, are to be turned around by regeneration process, and sustainable in the longer term.

Resourcing future regeneration

A major problem in regeneration policies has been a reduction in capital expenditure and a shift toward competitive bidding for available funds. Meanwhile, the nation's housing stock is ageing and clearly requires investment. Grant expenditure has declined even though renovation of existing stock is favoured over redevelopment.

There is a need to address other potential sources of funding, particularly in the private rented sector, many of which are explored by Leather and Younge (1998). Options for investment might include a decreased capital grant with loan charges for remaining costs, providing equity sharing loans, limiting of a grant to renewal areas, subsidised interest rate loans, an increased role for building societies, education, information, advice and practical help and reform of the building industry. This would need to be combined with an awareness of true costs of ownership and maintenance, a reduction in initial costs for first-time buyers, a change in attitude to borrowing, saving and insurance in the long-term, houses seen as liabilities as well as assets; use of housing equity for renovation works, a reduction in prices in run down areas, so releasing money for renovation, and targeting of grants to areas where there is no potential for private investment and other relevant assistance.

It remains to be seen what the local authority role will be longer-term in housing regeneration.

Summary

- The State has taken an active involvement in housing, and more recently wider urban regeneration, for many years, but particularly since the late 1960s in terms of private sector housing.
- More recent trends have been toward recognising the role of the wider environment in housing decline and a reflection of this in policy, made particularly explicit in Department of the Environment Circular 17/96: *Private Sector Renewal: A Strategic Approach* (DETR 1996).
- There has been a general shift from local to central government in decision-making on urban renewal seen through the funding opportunities and local authorities now bidding competitively for funding to preset criteria, resulting in new forms of management and local partnerships developing.
- Current policy seeks to take on board wider urban issues, to turn around problems such as social exclusion, so that regeneration can be sustainable in the longer-term, which is in line with academic research in the area.
- There is a need for local authorities constantly to appraise what they are doing and to be imaginative in how new organisations can become involved in the regeneration process, particularly in respect of securing increased resources.

5.6 Issues in regeneration – participation, partnerships and exclusion to inclusion

Outline

The relationship of poor housing and health, poverty and deprivation, lack of employment, poor local facilities and social exclusion has long been recognised, but it is only relatively recently that converging these areas has been reflected in policy. Until recently regeneration was dealt with under separate banners, but is now based on partnership working and a bringing together of all parties with potential skills, legislation and resources to move things forward. A comprehensive approach now seeks to draw together relevant personnel and organisations, both statutory and voluntary, to make a real and sustainable difference to urban renewal.

This new process seeks to engage with communities in a mutual learning process, where the professional also takes a new role – the regeneration agenda is no longer to be driven solely on a top-down basis. This new agenda seeks not only just to regenerate housing, but also the local environment, commerce and facilities so that inequalities can be addressed and local skills harnessed. There are many examples of good practice and the momentum local communities can bring to the process, if offered a real opportunity.

This section looks at community participation and empowerment in area regeneration, partnership working and combining these in moving from social exclusion to social inclusion.

Community participation and empowerment

Community participation has become a key concept in area regeneration policies, and there is much literature illustrating what it is about and how it can be encouraged to make policies more relevant, democratic and sustainable to local communities (e.g. Taylor 1995). Participation is essentially about power and knowledge. It encourages citizens to have a greater say in policy development, but it is the State that normally determines how much it will reorganise its power relationship with citizens. Participation plays a role in democracy and accountability.

Empowerment is about giving citizens greater input into policy-making and implementation by giving choice, so that the provider does not wholly determine outcomes. Empowerment gives the right to referendum, so existing decisions may be challenged and redetermined. An initial stage involves setting up community forums to decide how resources are used. The question of who is empowered is important and has implications for policy development and implementation. Is it the service consumer, the citizen or the whole community who are empowered? Or is the whole thing just lip service? Is the process genuinely designed to be inclusive, or in reality is it exclusive?

The spectrum of participation ranges from high levels of participation, where citizens have the authority to make decisions, to low levels, where citizens are given information after decisions are made. Participation is a dynamic process as relationships are constantly changing because of new or existing individuals or groups. The participation process can be thought of as a 'Ladder of Participation', where people have more say as they move up the ladder. At the bottom rung, citizens are simply told what is or has happened, whilst at the top, they share power with the relevant authority. The level they can get to depends on how much power and responsibility they are given, which can be influenced by the confidence citizens have in the State. Participation should be established if the State is serious about its commitment to local democracy. It helps develop sensitive and relevant policies, whereby marginal groups can gain a foothold into decision-making processes. If participation is positive and citizens feel they are being listened to, they are more likely to gain confidence and engage further. The process is resource-intensive and requires time, progress and review (Gaster and Taylor 1993).

Participation is about mutual learning. Citizens have 'grass roots' knowledge about how their communities work, whereas professionals have knowledge of policy processes. Citizens may need a lot of encouragement to become involved, because many have been isolated for far too long. Participation can bring variety, new ideas, relevant policies, inclusion of marginal groups and accountability into policy-making and implementation. It promotes a wider democracy. However, it can develop tensions if people do not feel heard or expect what they cannot be offered. It can be seen as 'lip service' only (Richardson 1983, Gaster and Taylor 1993).

Partnership working

Organisational change seems to be a constant feature of local government as new relationships continue to emerge in service delivery. The Welfare State required new forms of organisation to deliver services, which gradually evolved over the next decades. However, public sector organisation and management was criticised by proponents of the New Right and heavily cut through the 1980s and 1990s. The New Right argued that radical change was required to counteract perceived self-interest, white male-dominated management, dependency culture, and so on. To the New Right, the answer was to contract out and privatise services and introduce a market system into the public sector, with the notion and language of customer and of 'choice'. There was an increase in the voluntary sector as contractor, and the role of this sector tended to shift from policy advocate to mainstream provider. Alongside these ideologically driven changes came pressures from other sources. New social movements, including the disability lobby, ethnic minorities and women found a greater voice and were beginning to play an increasing role in policy development and implementation.

The notion of 'partnerships' has become an important concept in recent years. The idea of a partnership or multi-agency approach is a development and reworking of the above concepts. Partnerships are essentially about joint working with other organisations, which may be statutory or voluntary, public or private, large- or small-scale and may be formal or relatively informal in operation. Partnerships seek to engage personnel and organisations with similar objectives to provide better and more comprehensive services. They should be mutually supportive and enable a sharing of finances, information, commitment and other resources.

In terms of housing, examples of partnerships include working with private developers and registered social landlords to build more homes;

home improvement agencies; working with the police, local residents, social services, voluntary organisations to tackle issues such as crime; working with landlords and tenants to promote standards in the private rented sector; working with gas and electricity providers and HEES to improve domestic energy efficiency and so on (Goss and Blackaby 2000). They need to have sound leadership, with the right mixture of personnel at the relevant stages of the policy process, with good communication across all levels, so that effective and accountable progress can be made.

The change in environmental health and housing can be readily seen. There has been a marked shift from a bricks and mortar to partnership approach, both with individual clients and on an area basis. This change has been strongly encouraged by government policy through funding arrangements such as single regeneration budget, city challenge and estate action. Frequently the only way to secure such funding is to show 'accountability' through a partnership approach.

This is very much a part of bottom up rather than top down policy-making to prevent duplication, wasting resources, development of more sensitive and appropriate services which are all moving toward same end process, rather than defined product. The partnership approach may be very long term, such as turning around a poor housing estate, or over-seeing a renewal area as part of a social exclusion initiative. It is about bringing all players together to promote sustainable, longer-term change. The effects are likely to take months, if not years, to be felt.

Professional officers need to be part of that change, but there may be issues about who is overseeing the whole process. A partnership approach may or may not work, and it requires constant appraisal. It may be part of a long- or short-term project requiring input from different personnel at defined periods along the way. Considerable time and resources are required for partnerships to be successful.

Tackling social exclusion through regeneration

Social exclusion is a relatively new term and concept, originating in the European Union. It has a wider reference base than 'poverty', but poverty is one of its major causes and effects. Social exclusions is multi-faceted, including citizenship issues, relationships, powerlessness and low self-esteem that can result in generations of low-income households finding it difficult to aspire to access into mainstream society. Its remedy requires wide and innovative cross-organisational functions. For this reason, urban regeneration can play a key role in tackling social exclusion by starting to break down some barriers that prevent full inclusion into society through providing new opportunities and resources to deprived local communities. Such barriers are identified by London Voluntary Service Council (1998) as including:

- Discrimination.
- Poverty.
- Employment.
- Childcare.
- Fear of crime.
- Transport.
- Education.
- Health.
- Housing.

The Scottish Poverty Information Unit (1998) suggests that social exclusion and social integration should be seen in a dynamic perspective rather than a static one, linking macro- (such as social policies and the labour market), meso- (such as inner-city poverty) and micro-levels (structures of daily lifestyle of population). A key feature of social exclusion is that social mobility tends to be static and so poverty is continually reproduced. It is a complex and multifaceted issue linked inherently into work and unemployment, distribution of wealth and equity, racism, spatial distribution and urban management, as well as identity and political systems.

By definition, social inclusion must also exist and the manner in which people are included in society needs to be explored to find new and sustainable social models to develop. Social exclusion can be seen as a result of growing inequality from the 1980s, a rise in low-income household with more children living in poverty, higher rates of unemployment among young adults with an inability to find employment, pensioners on low incomes/benefit, difficulties in disabled people's access to society. In addition, disadvantage tends to be concentrated in certain communities (Joseph Rowntree Foundation 1999). The current Labour Government responded by establishing the Social Exclusion Unit in 1997.

The Chartered Institute of Housing (1998) argues for a holistic approach to regeneration, with housing as its cornerstone, and calls for a new and innovative approach. It suggested that the following key elements should be built into regeneration strategies to help combat social exclusion:

- Build on existing community strengths. The regeneration agenda should not be shaped by government fund allocation policies, but by communities themselves taking the lead in a bottom-up approach.
- Government policy should support neighbourhood initiatives, such as taxation, welfare reform and so on, which are beyond the control of local people.
- To explore and implement long-term, sustainable solutions, not short-term quick fixes and to maintain the momentum with adequate

revenue funding. Measures of success should have a long-term perspective.

- Resources should be carefully targeted to deprived areas, not just to high-profile examples.
- The emphasis should be on prevention to stop the decline before it happens through planning and creating mixed, sustainable communities with local facilities and services.
- To encourage joint working through relevant local partnerships to help develop and sustain a change in culture across government, agencies and organisations at the grass-roots level.

Current urban regeneration programmes seek to tackle areas, rather than problems in themselves, but in doing so invariably tackles some root issues in social exclusion. To be effective, such an approach needs to be very much developed by communities themselves as outlined above. Such a bottom-up approach policy can best support community initiatives. Long-term, sustainable solutions are required, not just high-profile examples and the emphasis needs to be on prevention. The whole process of turning communities from exclusion to inclusion requires a huge resource investment, innovation, commitment and energy from all involved if it is to be both successful and sustainable. Renewal areas are key to the process to tackling social exclusion in terms of private sector housing; much can be learned and reworked from successful parallel initiatives in the social housing sector.

Summary

- Community participation is key to the success and sustainability of any regeneration programme, so professionals can gain insight into the real rather than perceived needs and priorities of local people. Taking this on board at an early stage will encourage people to feel involved in the process, and will help sustain the greater input.
- Partnership working engages with a variety of local organisations, which also have a role in the regeneration process – regeneration is not now the sole local authority function. Again, this can bring new ideas and initiatives as well as important additional resources into the regeneration process.
- Turning areas and groups from social exclusion to inclusion takes considerable resources and initiative, but can at least be started through participation and partnership working, and one can work onwards and upwards from there.

Chapter 6

Conclusions

A huge amount of progress has been made in improving housing conditions since the mid-1800s, but there is still a long way to go. Many people still live in unsatisfactory or unfit housing, affecting their access to local amenities and community, whilst others have no home at all. Government ideology and funding opportunities have played a key role in housing provision and rehabilitation policies over this period, with capital funding providing the optimum way to provide and maintain decent housing, particularly to low-income groups. The rise and fall of the private rented sector has attracted controversy, but the main political parties are increasingly looking to this sector to meet the housing need, together with the opportunities and threats, which that sector provides. Many involved in enforcing conditions in the private sector will have their own views on the potential of the bottom end of the sector to meet the low-cost need to vulnerable groups.

The main factor in determining private sector housing renewal has to be that of resources. The history of housing renewal policy illustrates that the big quantitative (if not necessarily qualitative) leaps in improving housing conditions have tended to follow government financial investment cycles in housing. For example, the English House Condition Survey (DETR 1998) illustrated the major improvements in provision of internal amenities following grants such as improvement and intermediate grants from the 1970s targeted with investing in such facilities. Issues such as housing quality are harder to quantify, and a major defect of housing nationally is now disrepair, illustrating how housing conditions continue to change over time. As some housing is improved by a combination of owner's investment, grant-aid and other sources of funding, other housing is continually falling into disrepair. Local authorities alone have inadequate resources wholly to regulate and offer assistance when faced with continual decline in housing conditions, and have to look for alternative and innovative ways to help regulate an ageing private housing sector. There is the additional complication of private sector housing renewal of reliance on the owner to carry out renovation works and that person – owner, landlord or leaseholder – may be neither willing nor able to do so. This may have a substantial impact and effect on, for example, a wider area renewal scheme.

Housing strategies have become increasingly concerned with making optimum use of housing stock already there across all sectors, but in the private sector particularly in making use of the many thousand vacant properties that lie idle, whilst unacceptable numbers are recorded as homeless. Many local authorities have delivered vacant property strategies with some success, whilst others still have a long way to go. To be successful, the commitment of front-line officers needs to be backed up by financial commitment and support from local councillors and housing managers – and/or vice versa – with realistic resources. There is continued scope for local authorities to continue with what works well, but to approach new and review existing strategies to meet needs with energy and vision, taking on board legal requirements as well as being flexible and sensitive in their application.

There is continued growing awareness of conditions associated with HMOs and many local authorities have sound strategies and well-trained staff involved in delivering a service that is at the least complex, and at its worst – in more extreme cases – almost impossible given some of the irresponsible landlords who frequent this sector. There is a need for sound works in default and prosecution procedures. In some cases, tenants are also uncooperative due to the fear that they would be evicted or harassed, since many have no formal housing tenancy and are in a powerless position. Others may appear to be disinterested in their housing conditions, which can present difficulties for enforcement staff.

Aware of some of the issues unique to the private housing sector, the Chartered Institutes of Environmental Health and Housing have continued to campaign for decent housing, as well as publishing documents promoting uniformity in enforcement and training materials. There has been growing interest and membership of organisations such as Shelter, the National HMO Network, the Campaign for Bedsit Rights, the Bed and Breakfast Information Exchange and other Housing Forums to help promote standards. The continued pressure from such organisations has led to proposals to introduce a housing health and safety rating system and a national licensing scheme for HMOs, which feature in the current Housing Green Paper (DETR 2000).

The housing rating system involves a review of standards for living accommodation and a change of emphasis in their assessment to replace the statutory standard of fitness. It seeks to prioritise the worst dwellings first by differentiating between serious health and safety hazards and those where the overall risk to occupiers is marginal. It changes the emphasis from listing defects under currently defined standards, towards accounting for the effect of these defects so a rating can be applied. The rating approach would be a cumulative assessment and evaluation of health risks, informed by relevant scientific standards of the interrelationship of housing and health that could be based within existing local housing strategies. There seems to be a growing debate about the proposal and whether the existing statutory fitness standard is in fact the most practical and sensible way nationally to assess housing conditions – the proposals remain under consultation. In view of the additional risks in multiple occupied premises, there are also plans to license HMOs.

Many local authority housing officers have sought to use legislation available to them in innovative ways, including locally risk assessing unfit housing and registering, informally or otherwise, HMOs. More pioneering authorities have promoted housing and community issues through Local Agenda 21, an umbrella term for policies encouraging sustainable development. This has required a move away from thinking on statutory fitness toward a suitability for habitation and wider issues of sustainability. There has been much interest in the government's Social Exclusion Unit, and although we have yet to see what its outcomes will be, sound foundations are already there.

Recent housing renewal legislation has increasingly learned from and taken on board lessons from the past. Positive aspects of more recent legislation involves more emphasis on community involvement so that housing renewal programmes are more likely to meet need and, therefore, be sustainable as well as the potential for greater flexibility in some areas. The emphasis is inclined toward links to wider strategies, such as regeneration, anti-poverty and energy efficiency strategies, to optimise existing resources and reduce bureaucracy and overlap. However, it could also be argued that the new initiatives have created a new form of administrative bureaucracy rather that just having the required funding and other resources to be able to get on with the job.

If capital expenditure continues to decrease, then new viable ways forward – such as the potential for equity loans – rather than just waiting for the inevitable, need to be explored and made feasible. This may mean that the local authority's renewal role is designated to working in partnership with private sector financial institutions to secure loans on renovation works, or increased Agency working on a contracting basis to local authorities. Local authorities will need to account increasingly for value for money for expenditure on private sector housing renewal, and it seems likely that they will have to be increasingly innovative in maximising available capital spending. Whatever happens, local government is likely to be subject to increasing central controls over its spending and performance has become increasingly important in accessing available traditional and non-traditional forms of funding.

Crucially, long-term structural changes are necessary to deal with housing environments, more in line with the USA 1939 components for healthful housing cited in Section 2.2, with close cooperation with other health and welfare professionals. There is a need for sustainable strategies to improve physical structures of individual dwellings as well as wider social and environmental factors, and the chance to obtain housing that is safe and affordable. There is clearly a need for more low-cost social housing so that already disadvantaged households do not have to live in poor-quality private rented housing in the first place, which compounds their situation. However, the role of the environmental health service is bound by mainly reactive legislation that cannot address wider issues of disadvantage reflected in the housing system, which plays such a fundamental role in poor living conditions. Local authority officers can only enforce what is currently in legislation, not what is not there.

Despite many restrictions and frustrations, local authority officers concerned with promoting healthy housing have established forums to share knowledge and coordinate action, which has been beneficial in promoting standards of enforcement. This has helped some more reluctant officers and authorities take a greater interest in private sector housing conditions. Many housing forums are working well together to develop best practice and encourage councillors to take a more positive look at the private rented sector, targeting resources to areas of greatest need. Many local authority officers are working hard to improve housing conditions within the legislative framework available to them, continuing to have at least some impact in the poorest housing conditions to the most disadvantaged households.

Current key issues include developing and delivering increasingly flexible local housing strategies that are responsive to local need, that recognise and seek potential funding sources and are realistic about what they can hope to achieve. Continued pressure and campaigning can help raise awareness of housing conditions and, perhaps, encourage increased government expenditure in providing new and sustainable low-cost social housing as well as to maximise private sector potential by making optimum use of a huge resource that already exists.

References

Age Concern and RADAR (2000) *Report by Age Concern and Radar on the Disabled Facilities Grants System. Is the System Working?* http://www.radar.org.uk/information/campaign/disabledgrantfac.html (19 June 2000).

Albeson, R. (2000) 'Uneasy settlement', *Roof*, July/August, 16.

Arblaster, L. and Hawtin, M. (1993) *Health, Housing and Social Policy*, London: Socialist Health Association.

Ashton, J. and Seymore, H. (1988) *The New Public Health*, Milton Keynes: Open University Press.

Association of Metropolitan Authorities (1985) *Defects in Housing*, London: AMA.

Asylum Rights Campaign (1999) *Out of Sight, Out of Mind. A Report on the Dispersal of Asylum Seekers in the UK*. London: ARC.

Audit Commission (1991) *Audit Commission Local Government Report No. 6. Healthy Housing: The Role of Environmental Health Services*, London: HMSO.

Audit Commission (1992) *Audit Commission Local Government Report No. 9. Developing Local Authority Housing Strategies*, London: HMSO.

Balchin, P. (1995) *Housing Policy. An Introduction*, 3rd edn, London: Routledge.

Balchin, P., Bull, G. and Kieve, J. (1995) *Urban Land Economics and Public Policy*, London: Macmillan.

Balchin, P., Isaac, D. and Rhoden, M. (1998) 'Housing policy and finance', in P. Balchin and M. Rhoden (eds), *Housing: The Essential Foundations*, London: Routledge.

Barham, P. (1997) *Closing the Asylum: The Mental Patient in Modern Society*, Harmondsworth: Penguin.

Bassett, W. H. (1998) *Environmental Health Procedures*, 5th edn, London: E. & F. N. Spon.

Bassett, W. H. (ed.) (1999) *Clay's Handbook of Environmental Health*, 18th edn, London: E. & F. N. Spon.

Battersby, S. and Ormandy, D. (1999) 'Surveying the system', *Environmental Health Journal*, November, 356–359.

BBC (2000) *Health: Childhood Asthma Soars* http://news.bbc.co.uk/hi/english/health/newsid_363000/363403.stm (30 August 2000).

Boardman, B. (1991) *Fuel Poverty*, London: Belhaven.

Briggs, A. (1987) *A Social History of England*, 2nd edn, Harmondsworth: Penguin.

Brooks, E. (2000) 'Stoking the fire of low demand', *Housing Today*, 187 (8 June), 16–17.

Brown, C. and Savage, C. (1998) *For the Common Good. 150 years of Public Health: An Environmental Health Commemorative Issue*, London: CIEH.

Brown, G. W. and Harris, T. (1978) *Social Origins of Depression: A Study of Psychiatric Disorder in Women*, London: Tavistock.

Bullock, A., Stallybrass, O. and Trombley S. (1988) *The Fontana Dictionary of Modern Thought*, 2nd edn, London: Fontana.

CIEH (1994) *Amenity Standards for HMOs*, London: CIEH.

CIEH (1999) *Response to the Department of the Environment, Transport and the Regions Consultative Document on the Licensing of Houses in Multiple Occupation*, London: HMSO
http://www.cieh.org.uk/about/policy/response/hmocon.htm
(11 November 1999).

CIH (1996) *Housing Makes the Difference. The Chartered Institute of Housing's Programme for the Next Government*, Coventry: CIH.

CIH (1998) *Opening the Door: Housing's Essential Role in Tackling Social Exclusion*, Coventry: CIH.

CIH (2000) 'Asylum seekers', *Housing*, June (Policy in Practice Guide), 1–4.

Clapham, D., Kemp, P. and Smith, S. J. (1990) *Housing and Social Policy*, London: Macmillan.

Coleman, A. (1990) *Utopia on Trial: Vision and Reality in Planned Housing*, London: Hilary Shipman.

Conway, J. (ed.) (1988) *Prescription for Poor Health: The Crisis for Homeless Families*, London: London Food Commission, Maternity Alliance, SHAC and Shelter.

Conway, J. (2000) *Housing Policy*, Eastbourne: Gildredge.

Cowan, D. (1999) *Housing Law and Policy*, London: Macmillan.

Crisis (2000) *Health Action for Homeless People* http://www.crisis.org.uk (28 August 2000).

Cyberus (1999) *Sustainability* http://www.cyberus.ca/choose.sustain/Sustain.html (30 August 2000).

DETR (1996) *Housing Research Summary (No. 54, 1996). Private Landlords in England*, London: DETR http://www.housing.detr.gov.uk/hrs/hrs054.htm (3 August 2000).

DETR (1997) *Housing Research Summary (No. 56, 1997). Financing Temporary Accommodation in the Private Rented Sector: An Economic Analysis* http://www.housing.detr.gov.uk/hrs/hrs056.htm (30 August 2000).

DETR (1998a) *The English House Condition Survey 1996*, London: HMSO http://www.housing.detr.gov.uk/research/ehcs96/tables/index.htm (30 August 2000).

DETR (1998b) *Housing Fitness Standard: Consultation Paper*, London: HMSO.

DETR (1998c) *Modernising Local Government: Improving Local Services through Best Value*, London: HMSO.

DETR (1999a) *Fuel Poverty: The New HEES. A Programme for Warmer, Healthier Homes*, London: DETR.

DETR (1999b) *Allocation of Housing Capital Resources. Consultation Paper on Development of the Needs Indices Used in the Allocation of Housing Capital Resources to Local Authorities and Registered Social Landlords*, London: DETR.

DETR (1999c) *Licensing of Houses in Multiple Occupation: England Consultation Paper*, London: HMSO
http://www.housing.detr.gov.uk/information/consult/1hmoe/03.htm (11 November 1999).

DETR (1999d) *Housing Signpost (Issue 4) Summer 1999. Good Practice Guidance for Running Renewal Areas*, London: DETR
http://www.housing.detr.gov.uk/signpost/iss004/1.htm (11 July 2000).

DETR (1999e) *The Single Regeneration Budget, SRB Round 6*, London: DETR http://www.regeneration.detr.gov.uk/srb/index.htm (11 July 2000).

DETR (2000c) *Planning Policy Guidance Note No. 3: Housing. The Government's Response to Environment, Transport and Regional Affairs Committee's 17th report*, London: HMSO
http://www.planning.detr.gov.uk/response/ppg3/index.htm (4 September 2000).

DETR (2000a) *The Housing Green Paper: Quality and Choice: A Decent Home for All*, London: Department of the Environment, Transport and the Regions, and the Department of Social Security
http://www.housing.detr.gov.uk/information/consult/homes/index.htm (20 August 2000).

DETR (2000b) *Regeneration that Lasts*
http://www.housing.detr.gov.uk/research/lasts/6.htm (30 August 2000).

DETR and ENTEC UK (1997) *Fire Risk in HMOs. A Summary Report*, London: HMSO.

Deveraux, T. (1997) 'Group repair schemes', *Environmental Health, Private Sector Housing Monitor*, April.

DoE (1989) *Caravan Sites and Control of Development Act 1960 Section 5 Model Standards 1989 for Permanent Residential Mobile Homes Sites*, London: HMSO.

DoE (1992) *Circular 12/92: Guidance to Local Authorities on Standards of Fitness under Section 352 of the Housing Act 1985*, London: HMSO.

DoE (1993) *Local House Condition Survey Guidance Manual 1993*, London: HMSO.

DoE (1994) *Circular 18/94: Gypsy Sites Policy and Unauthorised Camping*, London: HMSO.

DoE (1994) *Inner Cities Research Programme: Assessing the Impact of Urban Policy*, London: HMSO.

DoE (1995) *Houses in Multiple Occupation: Establishing Effective Local Authority Strategies. Summary Report*, London: DoE.

DoE (1996) *Circular 17/97. Private Sector Renewal: A Strategic Approach.* London: HMSO.

DoE (1996) *Housing Research Summary No. 62 (1996): English House Condition Survey 1991 Energy Report*, London: HMSO.

DoE (1996) *Local Authority Houses in Multiple Occupation Survey 1995*, London: DoE.

DoH (1999) *Saving Lives: Our Healthier Nation*, London: HMSO. Online. Available http://www.ohn.gov.wk/ohn/ohn.htm (30 August 2000)

EHA (1998) *England's Empty Homes: National Breakdown of Empty Properties by Local Authority Area. Figures from DETR HIP 1 Returns Submitted by Local Authorities* http://www.emptyhomes.com/nstats.htm (1 September 2000).

EHA (1999) *Press Release on Empty Homes Bill*
http://www.emptyhomes.com (1 September 2000).

EHA (2000) *EHA Home Page* http://www.emptyhomes.com (1 September 2000).
Forrester, P. (1998) 'Environmental health and housing', in P. Balchin and M. Rhoden (eds), *Housing: The Essential Foundations*, London: Routledge.
Foskett, E. W. (1999) 'Historical development of environmental health in the UK', in W. H. Bassett (ed.), *Clay's Handbook of Environmental Health*, 18th edn, London: E. & F. N. Spon.
Fox, K. (2000) 'Families in temporary homes reach record high', *Inside Housing*, 15 September, 4.
Gaster, L. and Taylor, M. (1993) *Learning from Consumers and Citizens*, Luton: Local Government Management Board.
Gilliver, D. (2000) 'Media blitz', *Housing: Focus on Asylum Seekers*, June, 21.
Goss, S. and Blackaby, B. (1998) *Designing Local Housing Strategies: A Good Practice Guide*, Coventry: CIH and LGA.
Goss, S. and Blackaby, B. (1998) *Designing Local Housing Strategies*, CIH and LGA.
Hamnett, C. (1988) 'Conservative government housing policy in Britain, 1979–85; economics or ideology?', in W. van Vliet (ed.), *Housing Markets and Policies under Fiscal Austerity*, New York: Greenwood.
Hatchett, W. (2000) 'Action needed for asylum seekers', *Environmental Health News*, 28 January, 8.
Hatchett, W. (2000) 'Labour's search for an urban policy', *Environmental Health News*, 15(27), 9.
Hawkey, E. (2000) 'Lost in the system', *Housing, Focus on Asylum Seekers*, June, 22–23.
Hibbert, C. (1988) *The English: A Social History*, London: Paladin.
Home Office (1989) *Fire Statistics: United Kingdom*, London: HMSO.
Home Office (2000) *Government Statistics on Asylum Seekers* http://www.homeoffice.gov.uk/rds/pdfs/asy-jul00.pdf (4 September 2000).
Hooton, S. (1996) *A to Z of Housing Terminology*, Coventry: CIH.
Hutter, B. M. (1988) *The Reasonable Arm of the Law? The Enforcement Procedures of Environmental Health Officers*, Oxford: Clarendon.
Ineichen, B. (1993) *Homes and Health: How Housing and Health Interact*, London: E. & F. N. Spon.
Joseph Rowntree Foundation (1994) 'Filling England's empty homes', *Findings*, York: JRF.
Joseph Rowntree Foundation (1999) 'Housing and social exclusion', *Foundations for Housing*, March/April, York: JRF.
Leather, P. (1999) 'Housing conditions and housing renewal policy in the UK', in H. Skifter Andersen and P. Leather (eds), *Housing Renewal in Europe*, Bristol: University of Bristol and Policy Press.
Leather, P. and Mackintosh, S. (1993) 'Housing renewal in an era of mass home-ownership', in P. Malpass and R. Means (eds), *Implementing Housing Policy*, Buckingham: Open University Press.
Leather, P., Macintosh, S. and Rolfe, S. (1994) *Papering Over the Cracks: Housing Conditions and the Nation's Health*, London: National Housing Forum.
Leather, P. and Morrison, T. (1997) *The State of UK Housing: A Factfile on Dwelling Conditions*, York: Joseph Rowntree Foundation.
Leather, P. and Younge, S. (1998) *Repair and Maintenance in the Owner Occupied Sector*, Birmingham: University of Birmingham, School of Public Policy Centre for Urban and Regional Studies.

LGA (1999) *Guidance Note to Local Authorities in England and Wales Interim Arrangements for Asylum Seekers* http://www.lga.gov.uk/lga/asylum/index.htm (22 June 2000).

Local Government Management Board (1995) *A Framework for Local Sustainability*, Luton: LGMB.

London Voluntary Service Council (1998) *Barriers: Social and Economic Exclusion in London*, London: LVRC.

Lowry, S. (1991) *Housing and Health*, London: British Medical Journal.

LRC (2000) *BABIE – The Bed and Breakfast Information Exchange* http://www.london-research.gov.uk/hs/hsbabie.htm (30 August 2000).

Luba, J. (1991) *Repairs: Tenants' Rights*, 2nd edn, London: Legal Action Group.

Malpass, P. and Murie, A. (1999) *Housing Policy and Practice*, 5th edn, London: Macmillan.

Markus, T. A. (1993) 'Cold, condensation and housing poverty', in R. Burridge and D. Ormandy (eds), *Unhealthy Housing: Research, Remedies and Reform*, London: E. & F. N. Spon.

Moran, T. (1997) *Legal Competence in Environmental Health*, London: E. & F. N. Spon.

National Audit Office (1989) *Homelessness*, London: Auditor General.

National Consumer Council (1991) *Deathtrap Housing*, London: NCC.

National Energy Agency (1996) HECA 1995. Advice to Energy Conservation Authorities on Dealing with Households with Special Circumstances, Newcastle upon Tyne, unpublished.

National Federation of Housing Associations (1985) *Inquiry into British Housing Report*, London: NFHA.

Oatley, N. (1998a) 'Contemporary urban policy: summary of themes and prospects', in N. Oatley (ed.), *Cities, Economic Competition and Urban Policy*, London: Paul Chapman.

Oatley, N. (1998b) 'Cities, economic competition and urban policy', in N. Oatley (ed.), *Cities, Economic Competition and Urban Policy*, London: Paul Chapman.

Oatley, N. (1998c) 'Restructuring urban policy: the single regeneration budget and the challenge fund', in N. Oatley (ed.), *Cities, Economic Competition and Urban Policy*, London: Paul Chapman.

Ormandy, D. (1999) 'No fixed abode', *Environmental Health Journal*, April, 102.

Ormandy, D. and Burridge, R. (1988) *Environmental Health Standards in Housing*, London: Sweet & Maxwell.

Ormandy, D. and Burridge, R. (1993) 'The legal environment of housing conditions', in R. Burridge and D. Ormandy (eds), *Unhealthy Housing: Research, Remedies and Reform*, London: E. & F. N. Spon.

Oxby, R. (1999) 'Housing standards and enforcement', in W. H. Bassett (ed.), *Clay's Handbook of Environmental Health*, 18th edn, London: E. & F. N. Spon.

Page, J. (1996) Issues of housing and health relating to asylum seekers: draft report of the day of the conference held on 5 February 1996 jointly hosted by London Borough of Hackney Directorate of Environmental Health and the University of Greenwich School of Environmental Sciences. Unpublished report, London: London Borough of Hackney.

Ransom, R. (1991) *Healthy Housing*, London: E. & F. N. Spon.

Resource Information Service (2000) *Asylum Seekers and Refugees* http://www.homelesspages.org.uk/../subjects/S0000016.html (18 May 2000).

Rhoden, M. (1998) 'Policy-making and politics', in P. Balchin and M. Rhoden (eds), *Housing: The Essential Foundations*, London: Routledge.
Richardson, A. (1983) *Participation*, London: Routledge.
Richardson, K. and Corbishley, P. (1999) 'The characteristics of frequent movers', *Findings*, York: JRF.
Scottish Poverty Information Unit (1998) *Response to the Social Exclusion Consultation, SPIU Submission* http://spiu.gcal.ac.uk (12 October 1998).
Simpson, M. (2000) 'A blind spot in the vision', *Inside Housing*, 14 July, 11.
Skifter Andersen, H. (1999) 'Housing rehabilitation and urban renewal in Europe: a cross-national analysis of problems and policies', in H. Skifter Andersen and P. Leather (eds), *Housing Renewal in Europe*, Bristol: University of Bristol and Policy Press.
Skifter Andersen, H. and Leather, P. (eds) (1999) *Housing Renewal in Europe*, Bristol: University of Bristol and Policy Press.
Stewart, J. (1999) 'Healthy housing: the role of the environmental health officer', *Journal of the Royal Society for the Promotion of Health*, 199(4), 228–234.
Stewart, J. and Thompson, N. (1999) 'Living aboard. As safe as houses?', *Environmental Health Journal*, 107(5), 145–149.
Sustainability (2000) *Sustainability 2000 Home Page* http://www.sustainability2000.org/index_main.htm (30 August 2000).
Taylor, M. (1995) *Unleashing the Potential: Bringing Citizens to the Centre of Regeneration*, York: JRF.
Toulcher, P. (1998) 'Boosting the rating system', *Environmental Health Journal*, December, 364–365.
Townsend, P., Davidson, N. and Whitehead, M. (1992) *Inequalities in Health. The Black Report and the Health Divide*, London: Penguin.
United Nations (1992) *Earth Summit Press Summary of Agenda 21*, New York: Department of Public Relations.
WHO (2000) *About WHO* http://www.who.int/aboutwho/en/definition.html (30 August 2000).
Wilcox, S. (1998) *Housing Finance Review 1998/9*, York: Joseph Rowntree Foundation.
Winders, J. (1997) 'Adaptations for people with disabilities', *Environmental Health, Private Sector Housing Monitor*, March.

Further reading

Arblaster, L. and Hawtin, M. (1993) *Health, Housing and Social Policy*, London: Socialist Health Association.
Arden, A. and Hunter, C. (1992) *Manual of Housing Law*, 5th edn, London: Sweet & Maxwell.
Audit Commission (1991) *Audit Commission Local Government Report No. 6. Healthy Housing: The Role of Environmental Health Services*, London: HMSO.
Audit Commission (1992) *Audit Commission Local Government Report No. 9. Developing Local Authority Housing Strategies*, London: HMSO.
Balchin, P. (1995) *Housing Policy: An Introduction*, 3rd edn, London: Routledge.
Balchin, P. (1998) 'An overview of pre-Thatcherite housing policy', in P. Balchin and M. Rhoden (eds), *Housing: The Essential Foundations*, London: Routledge.

Bassett, W. H. (1998) *Environmental Health Procedures*, 5th edn, London: E. & F. N. Spon.
Bassett, W. H. (ed.) (1999) *Clay's Handbook of Environmental Health*, 18th edn, London: E. & F. N. Spon.
Bassett, W. H. and Rooney, R. M. (1999) 'Communicable disease, administration and control', in Bassett, W. H. (ed.), *Clay's Handbook of Environmental Health*, 18th edn, London: E. & F. N. Spon.
Blythe, A. (1999) 'Radon in buildings', in W. H. Bassett (ed.), *Clay's Handbook of Environmental Health*, 18th edn, London: E. & F. N. Spon.
Boardman, B. (1991) 'Fuel poverty. An energy efficiency solution to a social problem', *Environmental Health*, 100(3) 61–65.
Boardman, B. (1991) *Fuel Poverty*, London: Belhaven.
BRE (1989) *Information Paper IP 2/89: Improving Energy Efficiency in Housing*, Watford: BRE.
BRE (1991) *Housing Defects Reference Manual Defect Action Sheets*, London: E. & F. N. Spon.
BRE (1991) *Information Paper IP 15/91: Fire Spread between Caravans*, Watford: BRE.
BRE (1994) *Information Paper IP 11/94: Financial Benefits of Energy Efficiency to Housing Landlords*, Watford: BRE.
BRE (1998) *Digest 208 1998 Increasing the Fire Resistance of Existing Timber Floors*. CI/SfB(23)i(K2), Watford: BRE.
Brown, T. and Passmore, J. (1998) *Housing and Anti-Poverty Strategies*, Coventry, CIH and Joseph Rowntree Foundation.
Bryson, J. (1997) 'Managing Capital Programmes', *Environmental Health, Private Sector Housing Monitor*, October, i–iv.
BS 5266 *Code of Practice for Design and Installation of Emergency Lighting Systems in Dwellings*, London: BSI.
BS 5839 *Code of Practice for Design and Installation of Fire Detection and Alarm Systems in Dwellings*, London: BSI.
Building Research Establishment (1989) *Information Paper (IP 2/89). Improving Energy Efficiency in Housing*, Watford: BRE.
Burberry, P. (1983) *Environment and Services*, London: Mitchell.
Burry, P. (1996) *Fire Detection and Alarm Systems: A Guide to the BS Code: BS 5839: Part 1*, 2nd edn, Hertfordshire: Paramount.
Chudley, R. (1988) *Building Construction Handbook*, Oxford: Butterworth-Heinemann.
CIEH (1994) *Amenity Standards for HMOs*, London: CIEH.
CIEH (1995) *Fire Safety for Houses in Multiple Occupation. An Illustrated Guide by the Chartered Institute of Environmental Health*, London: Chadwick House Group.
CIEH (1995) *Travellers and Gypsies: An Alternative Strategy*, London: CIEH.
CIH (1996) *Energy Efficiency Good Practice Briefing Issue 6*, Coventry: CIH.
CIH (2000) *The Housing Green Paper. Quality and Choice: A Decent Home for All. A Summary and Initial Comments from the Chartered Institute of Housing*, Coventry: CIH.
Clapham, D., Kemp, P. and Smith, S. J. (1990) *Housing and Social Policy*, London: Macmillan.
Connelly, J. (1999) 'Housing conditions, diseases and injury', in W. H. Bassett

(ed.), *Clay's Handbook of Environmental Health*, 18th edn, London: E. & F. N. Spon.

Conway, J. (2000) *Housing Policy*, Eastbourne: Gildredge Social Policy.

Cowan, D. (1999) *Housing Law and Policy*, London: Macmillan.

Crisis (2000) *Health Action for Homeless People* http://www.crisis.org.uk/ (28 August 2000).

Critchley, R. (1991) *Fire Safety Guide for Multi-Occupied, Privately Rented Housing*, London: Shelter's Campaign for Bedsit Rights.

DETR (1996) *Housing Key Facts. Home Repair Assistance*, London: DETR.

DETR (1998) *Housing in England 1997/8*, London: HMSO.

DETR (1998) *Housing Research Summary: Controlling Minimum Standards in Existing Housing (No. 75, 1998)*, London: DETR http://www.housing.detr.gov.uk/hrs.hrs075.htm (17 February 1998).

DETR (1999) *English House Condition Survey 1996. Houses in Multiple Occupation in the Private Rented Sector*, London: DETR.

DETR (1999) *Towards an Urban Renaissance: Managing the Urban Environment*, London: DETR http://www.regeneration.detr.gov.uk/utf/renais/2.htm (11 July 2000).

DETR (2000) *Housing Health and Safety Rating System (HHSRS)*, London: DETR http://www.housing.detr.gov.uk/research/hhsrs/index/htm (14 September 2000).

DETR (2000) *Housing Research Summary No. 118, 2000: Collecting, Managing and Using Housing Stock Information: Good Practice Guidance*, London: DETR.

DETR (2000) *Housing Research Summary (No. 121, 2000). Report of the Park Homes Working Party* http://www.housing.detr.gov.uk/hrs.hrs121.htm (30 August 2000).

DETR (2000) *Housing Research Summary: Development of the Housing Health and Safety Rating System (No. 122, 2000)*, London: DETR http://www.housing.detr.gov.uk/hrs.hrs122.htm (19 July 2000).

DETR (2000) *Housing Research Summary: Housing Health and Safety Rating System: Quick Guide (No. 123, 2000)*, London: DETR http://www.housing.detr.gov.uk/hrs.hrs123.htm (19 July 2000)

DETR and Home Office (1998) *Joint Good Practice Guidance for Local Authorities and the Police on Managing Unauthorised Camping*, London: DETR and Home Office.

DoE (1992) *Circular 12/92: Guidance to Local Authorities on Standards of Fitness under Section 352 of the Housing Act 1985*, London: HMSO.

DoE (1993) *Circular 3/97: HMOs. Guidance on the Provisions in Part II of the Housing Act 1996*, London: HMSO.

DoE (1993) *Circular 12/93: HMOs. Guidance to Local Housing Authorities on Managing the Stock in their Area*, London: HMSO.

DoE (1993) *Local House Condition Survey Guidance Manual*, London: HMSO.

DoE (1994) *Circular 1/94: Gypsy Sites Policy and Unauthorised Camping*, London: HMSO.

DoE (1995) *Local Authority Houses in Multiple Occupation Survey 1995*, London: DoE.

DoE (1996) *Circular 2/96. Home Energy Conservation Act 1995*, London: HMSO.

DoE (1996) *Circular 17/96: Private Sector Renewal: A Strategic Approach*, London: HMSO.

DoE (1996) *Housing Research Summary No. 62 (1996): English House Condition Survey 1991 Energy Report*, London: HMSO.

DoE (1997) *Circular 5/97. Energy Conservation Act 1996*, London: HMSO.

DoE, Home Office and National Caravan Council (1989) *Fire Spread between Park Homes and Caravans*, London: HMSO.

Drew, R. and Thompson, K. (Royal Borough of Kensington and Chelsea) (1997) *Fire Precautions for Houses in Multiple Occupation: A Practical and Technical Guide*, London: Royal Borough of Kensington and Chelsea.

DTI (2000) *Home and Leisure Accident Report: Summary of 1998 data*, London: DTI.

Dyer, S. (1991) *DoE Mobile Homes in England and Wales. Report of a Postal Survey of Local Authorities*, London: HMSO.

EHA (1998) *England's Empty Homes: National Breakdown of Empty Properties by Local Authority Area. Figures from DETR HIP 1 Returns Submitted by Local Authorities* http://www.emptyhomes.com/nstats.htm (1 September 2000).

Everett, A. (1983) *Materials*, London: Mitchell.

Forrester, P. (1998) 'Environmental health and housing', in P. Balchin and M. Rhoden (eds), *Housing: The Essential Foundations*, London: E. & F. N. Spon.

Forshaw, R. (2000) 'Asylum plans will mean cash shortfall', *Housing Today*, April, 6.

Gabe, J. and Williams, P. (1993) 'Women, crowding and mental health', in R. Burridge and D. Ormandy, *Unhealthy Housing: Research, Remedies and Reform*, London: E. & F. N. Spon.

Gibson, T. (1979) *People Power. Community and Work Groups in Action*, Harmondsworth: Penguin.

Gordon, R. J. F. (1985) *The Law Relating to Mobile Homes and Caravans*, London: Shaw & Sons.

Goss, S. and Blackaby, B. (1998) *Designing Local Housing Strategies*, London: CIH and LGA.

Habgood, V. (1999) 'Pest control', in W. H. Bassett (ed.), *Clay's Handbook of Environmental Health*, 18th edn, London: E. & F. N. Spon.

Hall, S. and Held, D. (1989) 'Citizens and citizenship', in A. Hall and M. Jacques (eds), *New Times: The Changing Face of Politics in the 1990s*, London: Lawrence & Wishart.

Hatchett, W. (1999) 'HMO licensing fails to find a place in the next session', *Environmental Health News*, 14(15).

Hawes, D. and Perez, B. (1995) *The Gypsy and the State: The Ethnic Cleansing of British Society*, Bristol: SAUS.

Hinks, J. and Cook, G. (1997) *The Technology of Building Defects*, London: E. & F. N. Spon.

Hollis, M. (1988) *Surveying for Dilapidations: A Practical Guide to the Law and its Application*, London: Estates Gazette.

Home Office (1989) *Fire Statistics: United Kingdom*, London: HMSO.

Howarth, C., Kenway, P., Palmer, G. and Miorelli, R. (1999) *Monitoring Poverty and Social Exclusion: Labour's Inheritance*, York: Joseph Rowntree Foundation.

HSE (1986) *Guidance CS4: The Keeping of LPG in Cylinders and Similar Containers*, London: HSE

Hunter, C. (1998) 'Asylum seekers rights to housing: new recipients of old Poor Law', in F. Nicholson and P. Twomey (eds), *Current Issues of UK Asylum Law and Policy*, Aldershot: Ashgate.

Hutter, B. M. (1988) *The Reasonable Arm of the Law? The Enforcement Procedures of Environmental Health Officers*, Oxford: Clarendon.

Institution of Environmental Health Officers (1992) *Residential Moorings on Inland Waterways: A Professional Practice Note by the Institution of Environmental Health Officers*, Coventry: IEH and PHAS Publ.

Institution of Environmental Health Officers (1994) *The Licensing of Permanent Residential Mobile Homes Sites: A Professional Practice Note by the Institution of Environmental Health Officers*, London: IEH.

Jameson, H. (1998) 'Fighting fuel poverty', *Public Service and Local Government*, March.

Kenny, P. and Thorpe, H. (1997) *Mobile Homes: An Occupier's Guide*, London: Shelter.

Leather, P. and Macintosh, S. (1994) *The Future of Housing Renewal Policy*, Bristol: School for Advanced Urban Studies.

Leather, P. and Macintosh, S. (eds) (1994) *Encouraging Housing Maintenance in the Private Sector*, Bristol: University of Bristol, School for Advanced Urban Studies and Joseph Rowntree Foundation.

Leather, P. and Morrison, T. (1997) *The State of UK Housing: A Factfile on Dwelling Conditions*, Bristol: Policy Press and JRF.

Leather, P. and Younge, S. (1998) *Repair and Maintenance in the Owner Occupied Sector*, Birmingham: University of Birmingham, School of Public Policy Centre for Urban and Regional Studies.

Leather, P., Murie, A., Moyes, L. and Bishop, J. (1985) *Review of Home Improvement Agencies*, Bristol: SAUS, University of Bristol.

Letall, M. (1988) *Mobile Homes: An Occupier's Guide*, London: Shelter.

LGA (1999) *Immigration and Asylum Bill. LGA Briefing, February 1999* http://www.lga.gov.uk/lga/asylum/bill.htm (22 June 2000).

Line, B. (2000) 'A national housing policy? That's so "20th century"!', *Housing*, May, 18–20.

Luba, J. (1991) *Repairs: Tenants Rights*, London: Legal Action Group.

Macintosh, S. and Leather, P. (1993) *Renovation File: A Profile of Housing Conditions and Housing Renewal Policies in the UK*, Oxford: Anchor Housing Trust

Macphail, R. (1997) 'Meeting the HECA challenge head on', *Environmental Health*, 105(1), 19–21.

Malone, S. (2000) 'New Labour gets tough on refugees and asylum seekers', *A Monthly Marxist Review*
http://www.labournet.org.uk.so/18toughonasylum.thml (22 June 2000).

Malpass, P. and Murie, A. (1999) *Housing Policy and Practice*, 5th edn, London: Macmillan.

Mann, J. and Smith, A. (1990) *Who Says There's no Housing Problem? Facts and Figures on Housing and Homelessness*, London: Shelter.

Mason, D. (1995) 'Gypsies and travellers. New problems for councils', *Environmental Health*, 103(7), 149.

McManus, F. (1994) *Environmental Health Law*, London: Blackstone.

Moran, T. (1997) *Legal Competence in Environmental Health*, London: E. & F. N. Spon.

Mulgan, G. (1989) 'The power of the weak', in A. Hall and M. Jacques (eds), *New Times: The Changing Face of Politics in the 1990s*, London: Lawrence & Wishart.

National Federation of Housing Associations (1985) *Inquiry into British Housing Report*, London: NFHA.

NEA (1996) *HECA 1995: Advice to Energy Conservation Authorities on Dealing with Households with Special Circumstances*, Newcastle upon Tyne, NEA.

Nevin, B., Lee, P., Revell, K. and Leather, P. (forthcoming) *Managing Changing Housing Markets: Developing a Sub-regional Framework in North Staffordshire*, Birmingham: University of Birmingham, Centre for Urban and Regional Studies.

Niner, P. and Hedges, A. (1992) *DoE Mobile Homes in Survey*, London: HMSO.

Ormandy, D. (1992) *Professional Practice Note: The Law of Statutory Nuisance in Premises*, Coventry: IEHO and PHAS Publ.

Ormandy, D. and Burridge, R. (1988) *Environmental Health Standards in Housing*, London: Sweet & Maxwell.

Oxby, R. (1999) 'Housing standards and enforcement', in W. H. Bassett (ed.), *Clay's Handbook of Environmental Health*, 18th edn, London: E. & F. N. Spon.

Pawlowski, M. (1998) 'Legal studies, property and housing law', in P. Balchin and M. Rhoden (eds), *Housing: The Essential Foundations*, London: Routledge.

Preece, D. and Clapham, D. (2000) Raising damp and breaking mould, *Environmental Health Journal*, 108(1), 25–27.

Ransom, R. (1999) 'Health and housing', in W. H. Bassett (ed.), *Clay's Handbook of Environmental Health*, 18th edn, London: E. & F. N. Spon.

Rhoden, M. (1998) 'Equal opportunities and housing', in P. Balchin and M. Rhoden (eds), *Housing: The Essential Foundations*, London: Routledge.

Robertson, D. and McLaughlin, P. (1996) *Looking into Housing: A Practical Guide to Housing Research*, Coventry: CIH Housing Policy and Practice Series.

Skifter Andersen, H. (1999) 'The national context for housing renewal', in H. Skifter Andersen and P. Leather (eds), *Housing Renewal in Europe*, Bristol: University of Bristol and Policy Press.

Stewart, J. (1999) 'Healthy housing. The role of the Environmental Health Officer', *Journal of the Royal Society for the Promotion of Health*, 199(4), 228–234.

Stroud-Foster, J. (1983) *Mitchell's Structure and Fabric: Part 1*, London: Mitchell.

Stroud-Foster, J. and Harington, R. (1983) *Mitchell's Structure and Fabric: Part 2*, London: Mitchell.

Tenants Resource and Information Service (1992) *How to Inspect a House or Flat. A Do It Yourself Guide to Property Inspection for Tenants*, Newcastle upon Tyne: TRIS.

Toulcher, P. (1996) 'The Housing Act 1996', *Environmental Health* a (pull out guide), October.

Toulcher, P. (1998) Section 82 of the EPA: nuisance or salvation?, *Environmental Health Journal, Housing Monitor*, November.

Townsend, P., Davidson, N. and Whitehead, M. (1992) *Inequalities in Health: The Black Report and the Health Divide*, London: Penguin.

Trott, T. (1998) *Managing Private Rented Housing: A Good Practice Guide*, Coventry: CIH.

Wilcox, S. (1998) *Housing Finance Review 1998/9*, York: JRF.

Leaflets

British Waterways (1996) *Guidance Notes: Boat Safety Scheme.*
British Waterways (undated) *Environmental Health Guidelines for Residential Boats: Applying for a Social Fund Payment.*
British Waterways (undated) *Environmental Health Guidelines for Residential Boats: Guidance Notes.*
(all are available from British Waterways Customer Services, Willow Grange, Church Road, Watford WD17 4QA; tel.: 01923 201120).
DETR (1996) *A Better Deal for Tenants. Your New Right to Compensation for Improvements,* London: DETR.
DETR (1996) *Housing Key Facts. Assured and Assured Shorthold Tenancies: A Guide for Tenants,* London: DETR.
DETR (1996) *Housing Key Facts. Assured and Assured Shorthold Tenancies: A Guide for Landlords,* London: DETR.
DETR (1996) *Housing Key Facts. Disabled Facilities Grants,* London: DETR.
DETR (1996) *Housing Key Facts. Do You Rent, or Are You Thinking from Renting, from a Private Landlord?* London: DETR.
DETR (1996) *Housing Key Facts. He Wants Me Out. Protection Against Harassment and Illegal Eviction,* London: DETR.
DETR (1996) *Housing Key Facts. House Renovation Grants,* London: DETR.
DETR (1996) *Housing Key Facts. Letting Rooms in Your Home,* London: DETR.
DETR (1996) *Housing Key Facts. Letting Your Home is Now Easier and Safer,* London: DETR.
DETR (1996) *Housing Key Facts. Disabled Facilities Grants,* London: DETR.
DETR (1998) *Keep Your Home Free From Damp and Mould,* London: DETR.
DETR (2000) *Home Repair Assistance,* London: DETR.
DETR (2000) *Housing Key Facts. Regulated Tenancies,* London: DETR.
DETR (2000) *Mobile Homes: A Guide for Residents and Site Owners,* London: DETR.
DETR (2000) *Repairs: A Guide for Landlords and Tenants,* London: DETR.
(all are available from DETR Free Literature, PO Box 236, Wetherby, West Yorkshire, LS23 7NB; tel.: 0870 122 6236; fax.: 0870 122 62370).
Health and Safety Executive (January 1999) *Landlords. A Guide to Landlords' Duties: Gas Safety (Installation and Use) Regulations 1998,* London: HSE
Health and Safety Executive (December 1999) *Gas Appliances: Get Them Checked. Keep them Safe,* London: HSE.
(available from HSE Books, PO Box 1999, Sudbury, Suffolk CO10 6FS; tel.: 01787 881165; fax.: 01787 313995).
Health and Safety Executive Gas Safety home page http://www.open.gov.uk/hse/gas/index.htm.
Shelter's Campaign for Bedsit Rights (November 1998) *Fast Action on Repairs and Bad Conditions,* London: Shelter.
Shelter's Campaign for Bedsit Rights (November 1998) *Paying a Deposit,* London: Shelter.
Shelter's Campaign for Bedsit Rights (November 1998) *Rights to Repairs,* London: Shelter.
Shelter's Campaign for Bedsit Rights (April 1999) *Paying your Rent,* London: Shelter.

Shelter's Campaign for Bedsit Rights (March 2000) *Private Tenancies*, London: Shelter.

Shelter's Campaign for Bedsit Rights (March 2000) *Private Tenants' Rights to Fire Safety*, London: Shelter.

Shelter's Campaign for Bedsit Rights (undated) *Gas Safety*, London: Shelter.

Shelter's Campaign for Bedsit Rights (undated) *Harassment and Illegal Eviction*, London: Shelter.

Shelter's Campaign for Bedsit Rights (undated) *The HMO Management Regulations*, London: Shelter.

(all are available from Shelter's Campaign for Bedsit Rights, 88 Old Street, London EC1V 9HU; tel.: 020 7505 2135; fax.: 020 7505 2167).

Index

Lightning Source UK Ltd.
Milton Keynes UK
20 June 2010

155856UK00005B/8/P